HccH
HAGUE CONFERENCE ON
PRIVATE INTERNATIONAL LAW
CONFÉRENCE DE LA HAYE
DE DROIT INTERNATIONAL PRIVÉ

EXPLANATORY REPORT

ON THE

HAGUE CONVENTION ON THE LAW APPLICABLE TO CERTAIN RIGHTS IN RESPECT OF SECURITIES HELD WITH AN INTERMEDIARY

HAGUE SECURITIES CONVENTION

BY

ROY GOODE,
HIDEKI KANDA, AND
KARL KREUZER

WITH THE ASSISTANCE OF

CHRISTOPHE BERNASCONI
(PERMANENT BUREAU)

Edited by the | *Edité par le*

Permanent Bureau | *Bureau Permanent*
6, Scheveningseweg 2517 KT The Hague | *La Haye* The Netherlands | *Pays-Bas*

A C.I.P. Catalogue record for this book is available from the Library of Congress.

Printed on acid-free paper.

ISBN 90-04-14836-1

Published by Koninklijke Brill NV

Koninklijke Brill NV incorporates the imprints Brill Academic Publishers,
Martinus Nijhoff Publishers and VSP.

http://www.brill.nl

Printed and bound in The Netherlands.

TABLE OF CONTENTS

TABLE OF CONTENTS .... **v**

PART I: GENERAL INTRODUCTION **1**

I. History of the Hague project on securities held with an intermediary 3
II. The basic purpose of the Hague Securities Convention: updating and unifying conflict of laws principles to reflect market reality in securities holdings 8
A. Developments in commercial practices 8
1. From direct holdings to intermediated holdings 8
2. Immobilisation, dematerialisation and centralisation 10
B. Differences in legal systems 10
C. Providing appropriate conflict of laws rules for securities credited to a securities account 11
D. Examples 11
E. The conflict of laws analysis 16
1. The need for uniform conflict of laws rules 16
2. The traditional *lex rei sitae* principle 17
3. Firm and unambiguous rejection of the 'look-through approach' for conflict of laws purposes 18
4. The Place of the Relevant Intermediary Approach (PRIMA) 19
5. The basis for the Convention's primary rule: an express choice of law agreement between the account holder and the relevant intermediary 20
III. Key features of the Convention 21

PART II: COMMENTARY ON THE ARTICLES OF THE CONVENTION **27**

Chapter I DEFINITIONS AND SCOPE OF APPLICATION **29**

Article 1 Definitions and interpretation **29**

I. Definitions – Article 1(1) 30
II. Further definitional provisions 38
A. Disposition – Article 1(2) 38
B. Intermediary – Article 1(3) - (5) 40
1. Introduction 40
2. Article 1(3) 40
3. Article 1(4) 40
4. Article 1(5) 41
III. Additional terms used in this Explanatory Report 42

Article 2 Scope of the Convention and of the applicable law **44**

I. Introduction 45
II. Article 2(1): issues governed by the applicable law determined by the Convention – the provisions in detail 47
A. The nature of the list of issues 47
B. The content of the list 48

III. Article 2(2): law applicable to a disposition of or an interest in securities when the account holder's interest is a contractual claim against its intermediary 56
IV. Article 2(3)(a) and (b): Subject to Article 2(2), the Convention does not determine the applicable law to purely contractual or otherwise purely personal rights between the parties to an account agreement or a disposition................57
V. Article 2(3)(c) 58
Article 3 Internationality 59
I. Introduction 59
II. Illustrations of the applicability of the Convention 60
III. Applicability – Time 63
IV. Conflicts between territorial units within Multi-unit States 63
Chapter II APPLICABLE LAW 65
Article 4 Primary rule 65
I. Introduction 66
II. The primary rule: Article 4(1), first sentence 69
III. The Qualifying Office requirement: Article 4(1), second sentence, and Article 4(2) 72
A. Introduction 72
1. The conditions summarised 72
2. The time factor 77
3. "Office" 78
4. The qualifying activity 78
B. Detail of the provisions 79
1. Article 4(1)(a) 79
2. Article 4(1)(b) 80
3. Article 4(2) 80
IV. Article 4(3) 80
V. Dispositions by transfer from one securities account to another, including through a chain of intermediaries 82
Article 5 Fall-back rules 88
I. Introduction 89
II. The first fall-back rule: Article 5(1) 89
III. The second fall-back rule: Article 5(2) 90
IV. The third fall-back rule: Article 5(3) 91
V. Amendment of the account agreement 91
Article 6 Factors to be disregarded 92
Article 7 Protection of rights on change of the applicable law 93
I. Introduction 93
II. Article 7(1): Scope of the provision 96
III. Article 7(2): Definitions of 'old' and 'new' law 98
IV. Article 7(3) – General rule: Applicability of the 'new' law 98
V. Article 7(4) – Exceptions: Applicability of the 'old' law 99
VI. Article 7(5) – Priority issues 103
Article 8 Insolvency 106
I. Introduction 106
II. Scope of Article 8 as to insolvency proceedings 107

III. Article 8(1): Recognition of interests acquired prior to an insolvency proceeding ... 109
IV. Article 8(2): Effects of previously acquired interests in an insolvency proceeding ... 110
Chapter III GENERAL PROVISIONS ... 112
Article 9 General applicability of the Convention ... 112
Article 10 Exclusion of choice of law rules (renvoi) ... 113
Article 11 Public policy and internationally mandatory rules ... 114
I. Introduction ... 114
II. Article 11(1): Public policy exception ... 115
III. Article 11(2): Internationally mandatory rules of the forum ... 116
IV. Article 11(3): An important limitation ... 117
Article 12 Determination of the applicable law for Multi-unit States ... 120
I. Introduction ... 120
II. Article 12(1): The operation of the Convention's primary rule (Art. 4(1)) in relation to a Multi-unit State ... 121
III. Article 12(2)(a): The "law in force in a territorial unit of a Multi-unit State" 124
IV. Article 12(2)(b): Preservation of internal conflict of laws rules in respect of perfection by filing, recording or registration ... 124
V. Article 12(3): Possibility for Multi-unit States to make a declaration preserving internal conflict of laws rules in the context of Article 5 ... 126
VI. Article 12(4)): Option for a Multi-unit State to impose a geographically more stringent condition ... 128
Article 13 Uniform interpretation ... 131
Article 14 Review of practical operation of the Convention ... 132
Chapter IV TRANSITION PROVISIONS ... 133
Article 15 Priority between pre-Convention and post-Convention interests ... 133
Article 16 Pre-Convention account agreements and securities accounts ... 135
I. Introduction ... 136
II. Article 16(1) ... 137
III. Article 16(2) ... 137
A. Providing directions as to the application of Articles 16(3) and (4) ... 137
B. Declaration mechanism ... 138
IV. Article 16(3) ... 139
A. Interpretive rule ... 139
B. Declaration mechanism ... 140
V. Article 16(4) ... 141
VI. When applicable law is not determined using Article 16(3) or Article 16(4) . 143
Chapter V FINAL CLAUSES ... 146
Article 17 Signature, ratification, acceptance, approval or accession ... 146
Article 18 Regional Economic Integration Organisations ... 148
Article 19 Entry into force ... 150
I. Introduction ... 150
II. Entry into force of the Convention on the international plane (Art. 19(1)) ... 150
III. Entry into force of the Convention for a State, Regional Economic Integration Organisation or a territorial unit ... 151

A. For the States that bring the Convention into force (Art. 19(1)) 151
B. For subsequent States and for REIOs (Art. 19(2)) 151
Article 20 Multi-unit States **153**
Article 21 Reservations **154**
Article 22 Declarations **155**
Article 23 Denunciation **157**
Article 24 Notifications by the Depositary **158**

PART III: APPENDIXES **159**

APPENDIX 1 .... **161**

TEXT OF THE HAGUE SECURITIES CONVENTION **161**

APPENDIX 2 .... **173**

CHRONOLOGY OF THE NEGOTIATIONS LEADING TO THE HAGUE SECURITIES CONVENTION **173**

APPENDIX 3 .... **177**

LIST OF ALL THE PRELIMINARY DOCUMENTS PRODUCED DURING THE NEGOTIATIONS LEADING TO THE HAGUE SECURITIES CONVENTION **177**

APPENDIX 4 .... **179**

LIST OF ALL EXPERTS WHO PARTICIPATED IN THE DIPLOMATIC SESSION HELD FROM 2-13 DECEMBER 2002 .... **179**

INDEX BY SUBJECT **187**

PART I:
General Introduction

I. History of the Hague project on securities held with an intermediary[1]

Int-1 The history of the work relating to the Hague Convention on the *Law Applicable to Certain Rights in Respect of Securities held with an Intermediary* (hereinafter referred to as the Hague Securities Convention, the Securities Convention or the Convention)[2] goes back to a proposal made during the Special Commission on General Affairs and Policy of the Hague Conference on Private International Law, which met from 8-12 May 2000 in The Hague. During that meeting, Australia, the United Kingdom and the United States of America jointly suggested that the Hague Conference (hereinafter referred to as the HCCH) develop, in a fast track procedure, a Convention on the law applicable to interests in and dispositions of securities held with an intermediary. The need for uniform conflict of laws rules on a world-wide level had been the subject of intense discussion for nearly a decade by the international legal and financial communities.[3]

Int-2 As demonstrated by this discussion, the need for conflict of laws rules that reflect the reality of how securities are held, transferred and pledged today (*i.e.*, by electronic book-entry through securities accounts) has become critical as a result of the growth, speed and volume of cross-border securities transactions, which have been made possible by

---

1 The sequence of the most important intermediate steps of the project is recapitulated in Appendix 2 to this Report; in addition, a list of all the Preliminary Documents produced in relation to this project is contained in Appendix 3. The significance of the Preliminary Documents is discussed in para. Int-78.

2 The full text of the Convention is reproduced in Appendix 1 to this Report; each commentary on a specific Article (see Part II of this Report) also reproduces the text of the Article in question.

3 See in particular: Euroclear, *Cross-Border Clearance, Settlement and Custody: Beyond the G30 Recommendations*, Brussels 1993, p. 62; Randall Guynn, *Modernizing Legal Rules to Reduce Settlement Risk*, 1 Capital Markets Forum Yearbook, IBA 1993, pp. 172-179; Marc van der Haegen, *Transfer and Collateralization of Book-Entry Securities in Belgium*, IBA 1994; *Shaping the Global Back-Office: A Front-Office Concern*, Price Waterhouse 1994, p. 164; Robert Rice, *No certainties about securities*, Financial Times (21 November 1995), p. 15; Robert Rice, *Es urgente armonizar las legislaciones mundiales de valores*, Expansión (2 December 1995), p. VI; Randall D. Guynn, *Modernizing Securities Ownership, Transfer and Pledging Laws*, A Discussion Paper on the Need for International Harmonization, with Responding Comments by Professor James Steven Rogers (USA), Professor Kazuaki Sono (Japan) and Dr Jürgen Than (Germany), Capital Markets Forum, Section on Business Law, IBA 1996 (paper available in pdf format on the Internet at http://www.dpw.com/iba.modernization.pdf); Roy Goode, *The Nature and Transfer of Rights in Dematerialised and Immobilised Securities*, 2 Capital Markets Forum Yearbook, IBA 1996, pp. 399-421; Letter Endorsing IBA Law Reform Proposals from the International Society of Securities Administrators (22 July 1996) [now called International Securities Services Association]; Stéphane Mouy & Hubert de Vauplane, *La réforme du nantissement des titres dématérialisés*, Banque & Droit July-August 1996, pp. 3-7; Uniform Law Conference of Canada (ULCC), Tiered Holding System - Uniform Legislation Project, Report of the Production Committee, April 30, 1997, drawn up by Eric Spink (Reporter); Letter from the International Primary Market Association to Mr Marc Vereecken, Administrator, European Commission (22 May 1997); Letter from the International Swaps and Derivatives Association to Mr Marc Vereecken, Administrator, European Commission (22 May 1997); *The Oxford Colloquium on Collateral and Conflict of Laws*, A Special Supplement to Butterworth's Journal of International Banking and Financial Law, September 1998; Randall D. Guynn / Nancy J. Marchand, *Transfer of Pledge of Securities held through Depositories,* in: Hans van Houtte (ed.), The Law of Cross-Border Securities Transactions, London 1999; Dicey & Morris, *The Conflict of Laws*, ¶ 24-064 (13th ed. 2000).

technological advancements. The volume of trades and collateral transactions in OECD government and corporate securities, for example, has grown to nearly US$2 trillion ($ 2,000,000,000,000) or more per day. This means that the volume of these transactions exceeds the world's total GDP (approximately US$36 trillion in 2003[4]) every 18 trading days. If investors and credit providers cannot determine in advance with certainty what law governs the nature of their interests in securities transferred or pledged through intermediaries, they may be deterred from entering into many financially beneficial and risk-reducing transactions. This can depress the value of securities issued by private and governmental issuers, increasing their cost of capital. It can also artificially inflate the cost of credit in a particular country by making securities less valuable as collateral. In short, legal uncertainty as to the law governing the perfection, priority or finality of transfers and pledges imposes significant friction costs on ordinary transactions and operates as an important constraint on desirable reductions in credit and liquidity exposures.

Int-3 Such legal uncertainty can also cause liquidity to dry up very rapidly during times of financial stress. Financial institutions are increasingly likely to hold billions of dollars of gross claims against other institutions. Legal uncertainties tend to be exaggerated during times of distress. Persons exposed to unsecured credit risk tend to behave most desperately at such times. If there is uncertainty about whether the gross claims can be reduced through the liquidation of collateral to a small fraction of the gross amount, the institutions needing credit may not find a sufficient amount of credit available at any price at such times, unnecessarily increasing the systemic risk (*i.e.*, the risk that the failure of a major player will have a domino effect on the market as a whole) and proliferating the volume of bankruptcies. An unwillingness to make credit available or enter into new transactions with the troubled institutions could rapidly spread to other institutions, amplifying systemic risk.

Int-4 It is also no longer possible to insulate domestic issuers, investors and markets from these global market forces or related friction costs and risks. Local issuers, investors and markets are now part of a larger interdependent network that forms an increasingly globalised marketplace. The exploding growth in the value and velocity of global financial transactions will no longer safely permit a purely local focus.

Int-5 The deficiencies of traditional conflict of laws rules are well known among practicing lawyers who are called upon to give legal opinions as to the enforceability, priority or finality of various securities transfer or pledging arrangements. They have also attracted the attention of a growing number of financial policymakers and other observers.[5]

---

4 See World Data Profile, available at www.worldbank.org/data/quickreference/quickref.html (World Bank Group).

5 See Group of Thirty Study Group Report, Global Clearing and Settlement: *A Plan of Action – Twenty recommendations that constitute a plan of action for global clearing and settlement,* issued in January 2003 (http://www.group30.org/home.php); see in particular Recommendation 15.

The current gap between most conflict of laws rules and the circumstances to which they apply reflect outdated assumptions about how securities are held, transferred and pledged.[6]

Int-6 In light of these concerns, the Special Commission of May 2000 unanimously recommended that this topic be included, with priority, in the Conference's agenda for future work and that, without waiting for the next Diplomatic Conference, an Experts meeting be convened to examine, in collaboration with other international organisations and the private sector, the feasibility of drawing up a new instrument on this topic.[7] The necessity for adopting a fast track procedure for the project was justified by referring to the high volume of financial transactions, both in terms of number and value, that involve securities held with an intermediary; it was stressed that there was an immediate practical need to provide financial markets with legal certainty and predictability, in particular as to the law governing perfection requirements of dispositions of securities held with an intermediary.

Int-7 In preparation for the suggested Experts meeting, the Permanent Bureau had prepared a comprehensive Report, which identified and illustrated the most important questions relating to the topic and examined possible solutions.[8] The Experts meeting, which was held in January 2001 and chaired by Ms Kathryn Sabo (Canada), not only concluded that the HCCH should indeed draw up, on an accelerated basis, a new instrument in this field – the Drafting Committee, established by the meeting and chaired by Professor Sir Roy Goode (United Kingdom),[9] already produced a first draft of key provisions of the future Convention.

---

6 For an illustration of the deficiencies in the traditional conflict of laws rule relating to securities held with intermediaries, see the comparative law work edited by Richard Potok, *Cross Border Collateral: Legal Risk and the Conflict of Laws,* London 2002.

7 See the *Conclusions of the Special Commission of May 2000 on General Affairs and Policy of the Conference*, prepared by the Permanent Bureau, Prel. Doc. No 10 of June 2000, for the attention of the Nineteenth Session, pp. 25-26.

8 C. Bernasconi, *The Law Applicable to Dispositions of Securities Held Through Indirect Holding Systems,* Prel. Doc. No 1 of November 2000 for the attention of the Working Group of January 2001.

9 The Drafting Committee was composed of the following experts (if not indicated otherwise, the experts were members of the Drafting Committee for the entire duration of the project): Roy Goode, Chair (United Kingdom), Lars Afrell (Sweden), Olaf Christman (Germany) (from January 2001 until September 2002), Diego Devos (Belgium), Philippe Dupont (Luxembourg), Alexandre de Fontmichel (France) (from January 2002 until the Diplomatic Session in December 2002), Francisco Garcimartín Alferez (Spain), Hervé Le Guen (France) (from January 2001 until January 2002), Hideki Kanda (Japan), Karl Kreuzer (Germany), Guy Morton (United Kingdom), Maxime Paré (Canada) (from October 2001 until March 2002), Ulrik Rammeskow Bang-Pedersen (Denmark), Fabian Reuschle (Germany) (for the duration of the Diplomatic Session in December 2002) and Jim Rogers (United States), who was seconded by Joyce Hansen (United States) from May 2002 until the Diplomatic Session in December 2002. In addition, the Chair had also invited the following representatives of the International Bar Association to participate as international private sector experts in the work of the Drafting Committee: Randall Guynn (Davis Polk & Wardwell), and Antoine Maffei (De Pardieu Brocas & Maffei). The Permanent Bureau was represented by Christophe Bernasconi and by its Legal Advisor for the project, Richard Potok. Finally, the Drafting Committee benefited from the technical assistance of Michael Huber (Davis Polk & Wardwell).

Int-8 Following the January 2001 Experts meeting, the Permanent Bureau, assisted by the Legal Advisor for the project, Richard Potok (Australia), started to organise and supervise an innovative informal working process. In light of the technical and speedy nature of the project and the lack of previous work undertaken by the HCCH in this field, the Permanent Bureau, in particular Dr Christophe Bernasconi (First Secretary) who was put in charge of the work relating to the Securities Convention, closely cooperated with Mr Potok throughout the duration of the project. Mr Potok, who had been a member of the Australian delegation at the May 2000 meeting and, in that capacity, had made a presentation that greatly assisted the other Member States representatives to understand the nature of the project and to assess its practical importance, helped in particular to establish a dense network of industry representatives with whom to discuss practical aspects of the legal framework to be embodied in the Convention, and to organise and coordinate the informal working process. Such an informal working process was indeed felt to be necessary not only because of the very specialised nature of the subject matter, but in particular also with a view to responding to the need for an expedited completion of the Convention. This informal working process, which rapidly became a key feature of the project and without which the speedy completion of the negotiations would not have been possible, was characterised by four important factors: an active Drafting Committee that met frequently in between formal meetings to help the Permanent Bureau in submitting a series of intermediate preliminary drafts of the Convention, widespread consultations with Member States, continuous private sector industry participation, and transparency.

Int-9 The January 2001 Experts meeting had expressly mandated the Drafting Committee to continue its work after the meeting, with a view to further refining its first draft and to suggest additional drafting proposals. This mandate authorised the Drafting Committee not only to execute decisions taken in the plenary, but also to suggest new approaches and solutions that would assist in further building up consensus. The first of these informal meetings of the Drafting Committee was held in May 2001 in Paris (at the invitation of De Pardieu Brocas & Maffei). After the official endorsement of the securities project in June 2001 (see below para. Int-13), which explicitly included the approval of the informal working methods, the Drafting Committee's frequent informal meetings became a key feature of the negotiation process. Such meetings were subsequently held in Oxford (October 2001; at the invitation of the Chair of the Drafting Committee), Brussels (December 2001; at the invitation of the Euroclear Group), Frankfurt (March 2002; at the invitation of the German Central Bank), and London (May 2002; at the invitation of Davis Polk & Wardwell). These meetings allowed the Drafting Committee to reflect on all the comments made on previous drafts and to prepare new intermediate drafts of key provisions for the future Convention, which in turn would then be widely circulated by the Permanent Bureau for comments by Member States, industry representatives and any interested party. There is no doubt that these informal meetings of the Drafting Committee helped to significantly speed-up the process, while at the same time building consensus on all key provisions to be included in the final text of the Convention.

Int-10 Throughout the informal working process, the Permanent Bureau kept liaison with the official organs of the Member States of the HCCH to inform them about recent developments in the informal working process, discuss issues relating to the substance of the future Convention and invite them to submit comments that would allow the Drafting

Committee to take their suggestions into consideration. The intensive contacts with the Member States became particularly apparent when the Permanent Bureau organised two series of a total of 17 Regional Discussion Workshops (RDW) around the globe to discuss and assess with State officials, legal experts and market participants the most recent draft of the Convention available (these RDW were held in Copenhagen, Frankfurt (2x), Hong Kong, London, Milan, New York (2x), Paris (2x), Rio de Janeiro, Rome, Stockholm, Sydney, Tokyo, and Toronto (2x)).

Int-11 With a view to ensuring that the final Convention would meet practical market requirements, be effective and operational, the Permanent Bureau established early contacts with a number of private sector industry representatives around the globe (in particular lawyers in private practice and internal counsel) and kept liaison with them throughout the negotiations. Many industry representatives participated in the informal working process, including the RDWs mentioned above; in addition, several Member States and observers had industry representatives amongst their delegation at the formal meetings in The Hague. Also, some of these industry observers participated as experts in the work of the Drafting Committee (see footnote 9). This industry input was invaluable and helped to ensure that the final text of the Convention, in particular the primary conflict of laws provision (Art. 4), would indeed lead to the legal certainty and predictability the Convention was designed to provide.

Int-12 Most importantly, however, the entire informal working process was kept totally transparent by the Permanent Bureau's continuing production of explanatory Notes, charts and summaries reflecting all the opinions and views expressed during the consultation process, thus allowing all Member States (and indeed observers) not only to be informed of the latest status of the discussions, but also to react and submit new comments and suggestions. All the documents produced by the Permanent Bureau were uploaded on the HCCH-website, thus ensuring global visibility of the most current status of the project. There is no doubt that the Permanent Bureau's strict adherence to transparency was an essential ingredient of the successful working methods put in place for this project.

Int-13 In June 2001, during the first part of its Nineteenth Diplomatic Session of the HCCH, the Member States, among other things, officially endorsed the securities project, including the fast-track procedure and the informal working process. After intense informal preparations, a Special Commission, chaired by Ms Stefania Bariatti (Italy), was held in January 2002 in The Hague.

Int-14 Another important intermediate step in the negotiation process was reached in May 2002, when the Permanent Bureau, in accordance with the informal working process described above, took the initiative of suggesting a redraft of the core provisions of the future Convention, that is, the primary rule and the fall-back rules to determine the law applicable under the Convention. In particular, the redraft suggested that the Convention's conflict of law rules should no longer be based on any attempt to 'locate' a securities account or the office that maintains a securities account to determine the law applicable under the Convention. Instead, the Permanent Bureau suggested allowing the relevant intermediary and the account holder to agree on the applicable law and to submit this agreement to specific conditions.[10] This approach formed the basis for the primary rule (Art. 4) and the fall-back rules (Art. 5) that were eventually adopted by consensus.

Int-15 The final stage of the project was reached in December 2002, when the second part of the Nineteenth Diplomatic Session of the Hague Conference was held in the Hague to adopt the draft Securities Convention.[11] Thus, completion of this project took only a little more than two years and a half – a remarkably short period to develop an instrument that not only addresses difficult technical issues, but which also has to suit the needs and interests of business practice while at the same time taking into account the peculiarities of many different legal systems. One might also stress that this project was completed without the need for one single vote during the entire process. The Commission finalising the work on the Securities project during the Diplomatic Session was again chaired by Ms Stefania Bariatti (Italy). After a subsequent *toilettage* (clean up) of the text coordinated by the Permanent Bureau, the final text of the draft Convention was released in February 2003.

II. The basic purpose of the Hague Securities Convention: updating and unifying conflict of laws principles to reflect market reality in securities holdings

A. Developments in commercial practices

Int-16 Over the past half century there has been a marked change in the way in which shares, bonds and other investment securities are held, transferred and pledged. Two sets of commercial developments merit particular attention.

1. From direct holdings to intermediated holdings

Int-17 The first set of commercial developments is a movement from holding, transferring and pledging securities through the physical possession of security certificates or the recording of registered ownership or other interests on the issuer's books (*i.e.*, direct holding system), to intermediated holding systems where interests in the securities are held, transferred and pledged through book-entries to securities accounts. This movement was pioneered by the establishment of Central Securities Depositories (CSDs), where large quantities of securities of different issuers are immobilised or otherwise concentrated (see further comments in para. Int-21).[12]

---

10 *Proposal for a redraft of Articles 4 and 4 bis,* Prel. Doc. No 13 of May 2002, submitted by the Permanent Bureau.

11 A list of all experts who participated in the Diplomatic Session held from 2-13 December 2002 is reproduced in Appendix 4 to this Report.

12 The oldest CSD was the *Wiener Giro- und Cassenverein*, founded in 1872 in Austria, which was later followed by the German *Kassenverein* (now part of Clearstream International). The predecessor of Sicovam, the French CSD (now Euroclear France), was established during World War II. Plans of the New York Stock Exchange to set up a similar entity in the United States before World War I did not materialise; it was only after the paperwork crisis on Wall Street in the late 1960s that *The Depository Trust Company (DTC)* was established. The movement of establishing CSDs spread to almost every country in the world after the Group of Thirty published its landmark report in 1989 on domestic clearance and settlement systems (see *Clearance and Settlement in the World's Securities Markets,*

Int-18 In intermediated holding systems, there will be one or more intermediaries (a term deliberately defined very broadly) standing between the investor and the issuer. Typically, a CSD or ICSD will maintain securities accounts for a limited number of financial institutions, broker-dealers and other professional investors (these intermediaries are often called participants of the CSD or ICSD). These participants, in turn, maintain securities accounts for their customers, such as institutional or retail investors or further intermediaries, and so forth until an intermediary maintains a securities account for an ultimate investor. As a result, there may be a variable number of tiers between the investors at the bottom of the structure and the issuers at the top.[13] Moreover, in many systems, there is no record (*i.e.*, no identification or segregation) of an individual account holder's interest in the underlying securities at the level of the issuer's register or that of any intermediary other than the intermediary with whom the investor has a direct relationship. This serves to reduce costs and facilitate the efficiency of transfers.

Int-19 The legal approaches to intermediated holdings vary among legal systems. Under some legal systems, the investor's direct (relevant) intermediary has interests in securities corresponding to the investor's interest and those of the intermediary's other customers credited to a securities account with another intermediary and so on up the chain to the highest-tier intermediary (normally a CSD or ICSD); thus, in these legal systems, there is a separate interest at the level of each intermediary and the highest-tier intermediary (or its nominee) is the registered holder of the securities on the issuer's books. Under other legal systems, however, an intermediary acts as a mere record keeper of an investor's interest; thus, while the investor's interest results from a credit to a securities account and is transferable through book-entry, the credit of the securities to a securities account maintained by the investor's intermediary establishes a direct relationship between the investor and the issuer (as a consequence, the highest-tier intermediary is not registered on the issuer's book as the holder of the securities).

Int-20 It is important to stress that the Hague Securities Convention caters to the needs of both these approaches; it applies when securities are credited to a securities account, regardless of how the relevant substantive law classifies the nature of the right resulting from the credit of the securities to the securities account and independently of whether this right is against the investor's (*i.e.*, the relevant) intermediary, any other intermediary or the issuer (see further comments in paras. Int-22 *et seq.*, and, in particular, the commentary to Art. 2).

New York & London: Group of Thirty, March 1989). International Central Securities Depositories (ICSDs) allow for the immobilisation and concentration of securities issued by foreign issuers. Euroclear, Clearstream and SIS SegaInterSettle are widely recognised as ICSDs.

13 It should be noted that in some legal systems, in particular in Nordic States, investors are allowed to open their securities accounts directly with the local CSD (see further comments in paras. 1-7 and 1-36).

2. Immobilisation, dematerialisation and centralisation

Int-21 A related set of commercial developments is the immobilisation, dematerialisation and centralisation of securities through CSDs and ICSDs. The immobilisation of security certificates with CSDs and ICSDs or other intermediaries refers to the holding, transferring and pledging of interests in securities by book-entry to securities accounts without any change in possession of the underlying security certificates. It may be noted that the use of global or jumbo certificates to represent the entire issue of securities, instead of individual security certificates for each security, has increased as securities have been immobilised. Dematerialisation refers to the complete elimination of certificates to represent securities.[14] Dematerialised securities are represented solely by entries in records kept by authorised record-keepers and/or directly on the issuer's books. Centralisation refers to concentrating the bookkeeping of dematerialised securities and the safekeeping of immobilised securities through CSDs.

B. Differences in legal systems

Int-22 Although the same commercial developments have taken place to one degree or another in virtually every market in the world, different legal systems classify an account holder's rights resulting from a credit of securities to a securities account in rather different ways. In some systems, the account holder's rights are characterised as a regular deposit, special deposit, co-property rights in an identifiable pool of securities or some other form of property right traceable to individual securities. In many of these legal systems, even though one or more intermediaries stands between an account holder and the issuer, the intermediary has no legal significance and the account holder's rights are the functional equivalent of those of a direct owner. Thus, the account holder's rights may include the right to enforce the securities directly against the issuer, the account holder may be treated as the direct owner of the securities or the account holder may be permitted or required to be recorded as the registered owner on the issuer's books.

Int-23 In other legal systems, the account holder's rights are characterised as an irregular deposit, general deposit or some other form of a purely personal (contractual) right against the intermediary to the delivery or transfer of a given type and number of securities. In still other systems, the account holder's rights are characterised or denominated as the interest of a beneficiary under a trust, a statutory fiduciary interest, a *Gutschrift in Wertpapierrechnung*, co-property rights in a fungible, notional or book-entry pool of securities, security entitlements or some other bundle of property, contractual or other rights. In both of these legal systems, either the intermediary breaks the ownership chain between the account holder and the issuer or the intermediary is treated as the registered, legal or nominal owner of the securities and the account holders are limited to enforcing the securities indirectly against the issuers through their intermediaries.

---

14 France introduced a fully dematerialised system by adopting Art. 94-II of the Finance Law for 1982 (Loi de finances pour 1982) (Law N°81-1160 of 30 December 1981); Art. 94-II entered into force on 3 November 1984. US and many other government securities have been dematerialised for several decades; most corporate securities, however, continue to be issued in certificated form.

C. Providing appropriate conflict of laws rules for securities credited to a securities account

Int-24 The Convention is designed to apply in relation to all securities held with an intermediary, independent of how the rights resulting from a credit of securities to a securities account are classified by any legal system. Thus, as more fully discussed in the commentary to Article 2, the Convention applies to determine the law applicable to all the Article 2(1) issues in relation to all securities held with an intermediary, independent of whether they are treated as the functional equivalent of direct holdings, including a right to enforce the securities directly against the issuer or to be recorded as the registered owner on the issuer's books, or whether either the intermediary breaks the ownership chain between the account holder and the issuer or an intermediary is treated as the registered, legal or nominal owner of the securities and the account holders are limited to enforcing the securities indirectly against the issuer through their intermediaries. Thus, the decisive factor to determine whether or not the Convention applies in relation to any securities is that those securities have at any stage entered into the intermediated system by being credited to a securities account (see para. 1-16). If the securities are held directly, that is, are embodied in certificates in the possession of the investor or the investor's interest is reflected directly on the books of the issuer and there is no intermediary involved, the Convention does not apply. The key features of the Convention are set out in paras Int-49 *et seq.*

Int-25 As more fully discussed in the commentary to Article 2, the Convention applies to determine the law applicable to all the Article 2(1) issues in relation to all securities held with an intermediary, independent of whether an account holder's rights arising from a credit of securities to a securities account are classified as proprietary, contractual, mixed or other under any legal system. The Convention does not, however, determine the law applicable to purely contractual or other purely personal rights between an account holder and its intermediary *inter se* and between the parties to a disposition *inter se*, provided that the issue at stake is not one specified in Article 2(1) (see paras. 2-4 *et seq.*, 2-30, 2-32 and 2-33 on the relationship between Art. 2(1) and Arts. 2(3)(a) and (b)). For the purposes of the Convention each account has to be treated separately, the intermediary who maintains the account being referred to as "the relevant intermediary".

D. Examples

Int-26 The following example illustrates a typical intermediated holding pattern to which the Convention applies and identifies the relevant intermediary:

Example Int-1

Illinois Inc is incorporated in Illinois in the United States. Illinois Inc has 5,000,000 shares outstanding. They are all represented by a global certificate that is immobilised at The Depository Trust Company (DTC), a Central Securities Depositary (CSD) incorporated in New York. DTC keeps the actual global certificate representing the Illinois Inc shares in its vault in New York. Illinois Inc maintains a share register in Newark, New Jersey, with a registrar, NJ Registrar, located there. DTC's nominee, Cede & Co, is recorded in the share register as the registered owner of the 5,000,000 outstanding Illinois Inc shares.

Australian Investor has an interest in 50,000 Illinois Inc shares, which is reflected by a credit to a securities account entitled 'Australian Investor Securities Account' maintained for Australian Investor by its intermediary, French Bank, which is located in Paris.

French Bank, in turn, holds interests corresponding to Australian Investor's interest in the Illinois Inc shares, together with interests of other customers of French Bank, in the from of Illinois Inc shares credited to a securities account entitled 'French Bank Omnibus Customer Securities Account' maintained for French Bank by European ICSD organised in Belgium. In total, French Bank has 650,000 Illinois Inc shares credited to that securities account.

European ICSD, in turn, holds interests corresponding to French Bank's interest in the Illinois Inc shares, together with interests of other participants in the European ICSD, in the form of Illinois Inc shares credited to a securities account entitled 'European ICSD Omnibus Customer Securities Account' maintained for the European ICSD by New York Bank, the European ICSD's U.S. subcustodian. In total, European ICSD has 2,250,000 Illinois Inc shares credited to that securities account.

New York Bank, in turn, holds interests corresponding to European ICSD's interest in the Illinois Inc shares, together with interests of other customers of New York Bank, in the form of Illinois Inc shares credited to a securities account entitled 'New York Bank Omnibus Customer Securities Account' maintained for New York Bank by DTC. In total, New York Bank has 3,000,000 Illinois Inc shares credited to that securities account.

The relevant intermediary as regards Australian Investor is French Bank; as regards French Bank is European ICSD; as regards European ICSD is New York Bank; and as regards New York Bank is DTC. So each intermediary other than DTC is also an account holder in relation to its own intermediary.

Int-27 The following example illustrates another typical intermediated holding pattern to which the Convention applies:

Example Int-2

Takushima Inc is incorporated in Osaka (Japan). Takushima Inc has issued $ 5,000,000,000 aggregate principal amount of fully dematerialised debt securities (bonds) which are held through JASDEC, a Central Securities Depositary (CSD) incorporated in Tokyo.

Dutch Investor has $ 75,000,000 aggregate principal amount of the Takushima Inc bonds credited to a securities account entitled 'Dutch Investor Securities Account' maintained for it by AB-IN Bank, incorporated in Amsterdam (Netherlands).

AB-IN Bank, in turn, holds interests corresponding to Dutch Investor's interest in the Takushima Inc bonds, together with interests of other customers of AB-IN Bank in Takushima Inc bonds, credited to a securities account entitled 'AB-IN Bank Omnibus Customer Securities Account' maintained for AB-IN Bank by the Tokyo office of its subsidiary, AB-IN Bank International, incorporated under Hong Kong law. In total, AB-IN Bank has $ 350,000,000 aggregate principal amount of Takushima Inc bonds credited to that securities account.

The Tokyo office of AB-IN Bank International, in turn, holds interests corresponding to AB-IN Bank's interest in the Takushima Inc bonds, together with interests of other customers of the Tokyo office of AB-IN Bank International, credited to a securities account entitled 'AB-IN Bank Tokyo Omnibus Customer Securities Account' maintained for the Tokyo office of AB-IN Bank International by Nomura Securities, a participant in JASDEC. In total, the Tokyo office of AB-IN Bank International has $ 550,000,000 aggregate principal amount of Takushima Inc bonds credited to that securities account.

Nomura Securities, in turn, has a securities account maintained for it by JASDEC. This account, which is entitled 'Nomura Securities Omnibus Customer Securities Account', reflects the interests in respect of Takushima Inc bonds of all Nomura Securities' customers, including the Tokyo office of AB-IN Bank International. In total, Nomura Securities has $ 1,350,000,000 aggregate principal amount of Takushima Inc bonds credited to that securities account.

The relevant intermediary as regards Dutch investor is AB-IN Bank; as regards AB-IN Bank is AB-IN Bank International; as regards AB-IN Bank International is Nomura Securities; and as regards Nomura Securities is JASDEC.[15]

Int-28 For the sake of simplicity, the examples above involve only securities of a single company. The fact pattern occurring in practice, particularly where portfolios of securities are provided as collateral, will often be much more complicated.

Int-29 Within intermediated holding systems, transfers of securities held with an intermediary occur by book entries without any form of delivery of certificates. In order to effect a transfer between two investors who are customers of a single intermediary, it is sufficient to make debit and credit entries in the books of that intermediary and, in some instances, forward information about the bottom-tier account holder to the issuer.

Int-30 In the case where two investors hold their interests in respect of securities through different intermediaries (a very likely scenario), the transfer may be slightly more

15 Although not applicable to the situation described above, under some legal systems, each intermediary standing between a bottom-tier account holder and the issuer may be required to forward the bottom-tier account holder's name and position in the securities to the issuer to be recorded as the registered owner on the issuer's books. Because the interests in the securities result from the credit of the securities to an account maintained by an intermediary, the Convention determines the law applicable to all the Article 2(1) issues in relation to such interests, even though the account holder is recorded as the registered owner on the issuer's books.

complicated. The fundamental principle is that there is only one relevant intermediary for a particular securities account.[16] Thus, to effect a disposition from a transferor to its ultimate counterparty (transferee) in intermediated holding systems, entries would typically be made to the transferor's securities account maintained by its relevant intermediary, to a securities account of that intermediary maintained by that intermediary's relevant intermediary, to a securities account maintained by the latter intermediary for the transferee's relevant intermediary, and, finally, to the securities account of the transferee maintained by its relevant intermediary. The conflict of laws provision of the Convention will not assure, or even typically produce, the result that one single law will govern all the issues specified in Article 2(1) with respect to credits and dispositions at all levels in the chain of intermediaries and account holders between the originating account holder and the ultimate counterparty. As a result, the crucial question whether or not the transferee acquires an interest in securities credited to its securities account free of adverse claims is governed by the law determined by the Convention with respect to the transferee's securities account.[17] If, under that law, the transferee acquires such an interest, the questions whether the corresponding debit of securities to the transferor's securities account or the corresponding credit of securities to the securities account of each intermediary standing between the transferor and the transferee were effective become irrelevant to the transferee (absent contrary contractual provisions). If, on the other hand, under the Convention law determined with respect to the transferee's account, the transferee's title is dependent on that of its relevant intermediary (or the transferor or any other intermediary in the chain between the transferor and the transferee), a court applying that law will look to the Convention law determined with respect to the intermediary's (or such other person's) securities account with its intermediary to determine what sort of title the transferee acquired under the Convention law determined with respect to the transferee's securities account.

Int-31 The following example illustrates a transfer by way of book entries between two investors who hold their interests in respect of securities through different intermediaries:

Example Int-3

> Australian Investor, who has a securities account maintained for it by French Bank, wants to borrow money and to provide as collateral for the borrowing its interest in

---

16 Application of the Convention's conflict rules in the context of a disposition of securities effectuated by transfer from one securities account to another, including through a chain of intermediaries, is discussed at length and illustrated by useful examples in paras. 4-43 *et seq.*

17 As more fully discussed in paras. 4-43 *et seq.* (see in particular para. 4-49), intermediaries manage the risk of being required to honour the credit of securities to accounts maintained for their account holders when a corresponding credit of securities to their accounts with their intermediaries is not honoured or is reversed (the risk of double liability) in a variety of ways. They may reallocate this risk by bargaining for the contractual right to reverse any credit of securities made to accounts maintained by them if a corresponding credit is not made or is reversed to accounts maintained for them by their intermediaries, spread the risk over a defined pool of their account holders, or consciously decide to assume and manage the risk.

respect of 500,000 Illinois Inc Shares. It seeks a loan from London Bank, an international investment bank incorporated in the UK and based in London.

London Bank has a securities account maintained for it by Swiss Bank, which is incorporated in Switzerland and has its principal place of business in Zurich. Swiss Bank has a securities account entitled 'Swiss Bank Omnibus Customer Securities Account' maintained for it by European ICSD.

London Bank and Australian Investor enter into a transfer of title by way of security agreement. London Bank instructs Australian Investor to transfer the Illinois Inc shares to London Bank's account with Swiss Bank (by requiring the shares to be moved to an intermediary that London Bank trusts, London Bank is avoiding any exposure to French Bank, in particular, to the possible consequences of administrative error, wrongdoing or insolvency of French Bank or the likely influence that Australian Investor may have with French Bank).

For the original account holder (Australian Investor), the relevant intermediary is French Bank, that is, the intermediary that maintains Australian Investor's securities account and with which Australian Investor has entered into the account agreement. For French Bank, the relevant intermediary is European ICSD, that is, the intermediary that maintains French Bank's securities account and with which French Bank has entered into the account agreement. Thus, French Bank, in addition to acting as an intermediary for Australian Investor, is also an account holder of European ICSD. For the ultimate transferee (London Bank), the relevant intermediary is Swiss Bank, that is, the intermediary that maintains London Bank's securities account and with which London Bank has entered into the account agreement. For Swiss Bank, the relevant intermediary is European ICSD, that is, the intermediary that maintains Swiss Bank's securities account and with which Swiss Bank has entered into the account agreement. Thus, Swiss Bank, in addition to acting as an intermediary for London Bank, is also an account holder of European ICSD. European ICSD is the common intermediary for both French Bank and Swiss Bank. Although the chain includes multiple intermediaries and, thus, multiple account holders, there is only one relevant intermediary for each account holder.

To effect a transfer of title from Australian Investor to its counterparty London Bank, entries would typically be made to:

(i) Australian Investor's securities account maintained for it by French Bank;

(ii) French Bank's securities account maintained for it by European ICSD;

(iii) Swiss Bank's securities account maintained for it by European ICSD; and

(iv) London Bank's securities account maintained for it by Swiss Bank.

As a result, the question whether London Bank acquires an interest in the securities credited to the securities account maintained for it by Swiss Bank free of all adverse claims is governed by the Convention law determined with respect to that securities account (even if Swiss Bank has the contractual right to reverse the credit if the corresponding credit of securities to Swiss Bank's securities account maintained for it by European ICSD is not effective or is reversed). If under that law London Bank takes the securities free of all adverse claims, the questions whether the corresponding debit of securities to Australian Investor's securities account or the corresponding credits of securities to French Bank's and Swiss Bank's respective securities accounts maintained by European ICSD were effective become irrelevant to London Bank (unless as a separate contractual matter, London Bank has agreed that Swiss Bank may reverse the credit if the corresponding credit of securities to Swiss Bank's account is not effective or is reversed). If, on the other hand, under the Convention law determined with respect to London Bank's securities account, London Bank does not acquire rights in the securities greater than Swiss Bank (or the Australian Investor or French Bank) had to convey under the Convention law determined with respect to Swiss Bank's (or the Australian Investor's or French Bank's) account, a court applying the Convention law determined with respect to London Bank's account will look to the Convention law determined with respect to Swiss Bank's securities account with European ICSD (or Australian Investor's account with French Bank or French Bank's account with European ICSD) to determine what sort of rights Swiss Bank (or Australian Investor or French Bank) had to convey to London Bank (on transfers across several intermediaries, see the further comments in paras. 4-43 *et seq.*).

Int-32 Today, the vast majority of securities around the world are held with intermediaries. The system caters to the needs of a global marketplace: it reduces the processing and settlement costs and risks of loss, theft and counterfeiting associated with a direct holding system. The transfer of interests in securities by mere accounting entries on the books of one or more intermediaries also allows for a rapid and efficient disposition of those interests, both domestically and across borders. As a result, the settlement time frames have shrunk.

E. The conflict of laws analysis

1. The need for uniform conflict of laws rules

Int-33 Prior to the Hague Securities Convention there was no uniform conflict of laws rule to govern issues that are of crucial practical importance for holdings and dispositions of securities held with an intermediary, in particular their effects against the intermediary and third parties. Thus, determination of the law applicable to issues such as the nature of the account holder's entitlement or the nature of a disposition between a collateral provider and a collateral taker depended on the choice of forum and the vagaries of whatever law might be found to apply.

Int-34 Uncertainties as to the approach to be adopted for determining the applicable law lead to significant expense for market participants (including collateral providers, because the expense of investigating perfection requirements is likely to be passed on to them by collateral takers). Moreover, because the position in many cases could not be satisfactorily determined, there remained an element of risk, which, given the aggregate size of transactions involved (nearly US$2 trillion per day and growing, see para. Int-2) and the identity and importance of the relevant financial institutions, had prudently to be regarded as systemic in character. Accordingly, agreement on a convention that would produce a uniform and rational resolution of these issues was likely to bring very important benefits to market users, market participants and the financial system as a whole. In particular, a person taking a security interest in securities held with an intermediary needs to know what law will govern the creation, perfection and priority of its interest. This *ex ante* legal certainty is essential for the smooth operation of the financial markets.

Int-35 The Hague Securities Convention is a pure conflict of laws convention. Thus, it does not prescribe any rules of substantive law governing interests in securities and allows the differences among legal systems in this regard to continue to co-exist (unity in diversity).[18] But why was there a need to develop a new conflict of laws rule? Why are the traditional conflict of laws approaches not adequate for the issues falling within the Convention's scope? How did the content of the conflict of laws rules introduced by the Convention develop during the negotiations? And what is the content of the Convention's primary conflict of laws provision in its final form?

2. The traditional *lex rei sitae* principle

Int-36 The traditional rule for determining the enforceability of a transfer of property (whether outright or by way of security) or any grant of a security interest (whether possessory or non-possessory) effected in the direct holding system is the *lex rei sitae* (or the *lex situs*), more specifically referred to in the context of securities as the *lex cartae sitae*. Under this rule, the effectiveness of a transfer of securities is determined by the law of the place where the securities are located at the time of the transfer. In the case of bearer securities (*i.e.*, securities which are represented solely by physical certificates or other documents of title and which are not registered in any name), this is taken to be the law of the place of the certificates representing the securities at the time of transfer. In the case of registered securities, the *lex rei sitae* is either the law of the place of the issuer's incorporation or organisation or the law of the place where the register is maintained at the time of transfer.

---

18 UNIDROIT has set up a Study Group to examine the possibility of harmonising security interests in investment securities as part of a wider capital markets project called *Transactions on transnational and connected capital markets - Study LXXVIII* (see the *Draft convention on substantive rules regarding securities held with an intermediary* (Preliminary Discussion Draft), submitted in April 2004 (available under http://www.unidroit.org). Within the Organisation of American States (OAS), it was suggested that the development of new uniform rules on cross-border investment securities transactions be added to the topics for the agenda of the Seventh Inter-American Specialized Conference on Private International law (CIDIP-VII) (see under http://www.oas.org). The European Commission has also started work in this field, see the Communication from the Commission to the Council and the European Parliament: Clearing and Settlement in the European Union – The way forward, adopted on 28 April 2004 (see under http://europa.eu.int/comm/internal_market/financial-markets/index_en.htm; in the Communication, see in particular the specific references to the Legal Certainty Project).

Int-37 These traditional approaches have generally produced a clear and satisfactory result in relation to direct holdings of securities. They are, however, unsatisfactory in relation to holding systems that involve intermediaries as they would require an approach that, for purposes of determining the applicable law, 'looks through' tiers of intermediaries to the level of the issuer, register or actual certificates (the 'look-through' approach). Such a 'look-through' approach gives rise to severe difficulties in relation to securities held with intermediaries, particularly if the register is not easily identifiable with a single location (*e.g.*, when it is electronic and serviced by more than one office).

3. Firm and unambiguous rejection of the 'look-through approach' for conflict of laws purposes

Int-38 For conflict of laws purposes, the Convention firmly rejects the 'look-through' approach. This is because in today's markets, it is very common for investors to provide collateral in the form of diversified portfolios of securities issued by companies organised under the laws of a number of different jurisdictions or by various government and governmental agencies, a practice greatly facilitated by modern movables security laws. Rules based on a 'look through' approach compel the collateral taker to determine and satisfy the perfection requirements of the domestic movables security law for each of the States where the issuers of the particular debt or equity securities are organised, a multitude of States for each such transaction. In the situation where the portfolio is not static but changes composition over time or even daily or hourly (again, a practice greatly facilitated by modern movables security laws), it is impossible, under such conflict of laws rules, for the collateral taker efficiently to manage a security interest in the portfolio. In addition, in many jurisdictions it is uncertain exactly which of the rules based on a 'look through' approach should be applied: the law of the place of the issuer, the place of the securities register, the place of the underlying security certificates, the law of the highest-tier intermediary or any other intermediary? Finally, even if the collateral taker did know which test to apply, it is often not possible to obtain the information necessary to apply the test. For example, an investor holding securities through various tiers of intermediaries may not be able to discover where the certificates are actually stored.

Int-39 In relation to many intermediated holding systems, conflict rules based on a 'look through' approach can also present other conceptual and practical problems. Where interests in securities are held in a fungible ('omnibus') account, there is typically no record of an individual account holder's interest in the underlying securities at the level of the issuer's register or that of any intermediary other than the intermediary with whom the investor has a direct relationship. Thus, if a person (*e.g.*, a collateral taker or a levying creditor), asserting a claim against the investor's interest in respect of the securities, were to try to enforce that claim at any of the higher levels in the multi-tiered holding system (with the record owner (*e.g.*, a CSD) being the top tier and the investor's relationship with its intermediary being the bottom tier), the response would typically be that at each upper-tier level no record exists of any interest of the investor against which the claim could be pursued. Moreover, the difficulty of matching an intermediary's receipt of securities to any securities credited to a particular customer's account is exacerbated by the effect of multilateral netting in clearing systems.

Int-40 One should add that the rejection of the look-through approach for conflict of laws purposes has no impact on any 'look-through' that may be necessary for substantive law purposes. As already mentioned (see comments in para. Int-22), in some systems, the account holder's rights resulting from the credit of securities to a securities account are characterised as a form of property right traceable to individual securities. In many of these legal systems intermediaries standing between an account holder and the issuer have no legal significance and the account holder's rights are the functional equivalent of those of a direct owner. Thus, the account holder's rights may include the right to enforce the securities directly against the issuer, the account holder may be treated as the direct owner of the securities or the account holder may be permitted or required to be recorded as the registered owner on the issuer's books.

4. The Place of the Relevant Intermediary Approach (PRIMA)

Int-41 Against the background of these elementary difficulties of the 'look-through' approach from a conflict of laws perspective, it was initially suggested that the connecting factor of the Convention's primary conflict of laws rule should focus on the location of the account to which the interest in the securities is credited, that is, the account maintained by the investor's immediate – or relevant – intermediary or the location of the office maintaining such account. This concept was described as Place of the Relevant Intermediary Approach (PRIMA). It was designed to reflect the fundamental principle that, in relation to securities held with an intermediary, the law applicable to the effects against the intermediary and third parties of holdings and dispositions of such securities is determined by the place of the office of the intermediary with which the account is maintained, without consideration of any higher-tier intermediary or the issuer of a security, that is, without 'looking-through' the chain of intermediaries.

Int-42 The major advantage of this approach is that it subjects all of an account holder's interest with respect to a portfolio of securities, and all dealings with that interest, to the law of one single jurisdiction, even where issuers, registers, certificates evidencing the underlying securities, or any higher-tier intermediaries are situated in different countries.

Int-43 The reference to the place of the relevant intermediary suggested that a single situs of the relevant intermediary or the securities account it maintains for the account holder had to be determined. The history of the negotiations leading to the Convention clearly revealed, however, that there is no criterion – generally acceptable on a global basis for the vast majority of transactions – to precisely and unequivocally determine the location of a securities account or the office of an intermediary that maintains a specific securities account. Particular consideration was given to the possibility of using tax, regulatory reporting or accounting requirements to determine the location of a securities account or the office maintaining a securities account; in certain States, intermediaries may be required to assign a code to each securities account that effectively allocates it to a particular office for tax, regulatory reporting or accounting purposes. This approach was eventually unanimously rejected. First, such tax, regulatory reporting or accounting requirements are by no means universal and may be the exception rather than the rule. Some States have comprehensive accounting rules for assets and liabilities that appear on an intermediary's balance sheet, but many do not have such rules for securities accounts that do not appear on the intermediary's balance sheet. Secondly, accounting, regulatory, and tax rules are based on considerations that are not necessarily related to the private law issues involving the global business of securities custody, clearing and settlement. Consequently, to use the allocation of a securities account to a particular office for tax, regulatory reporting or accounting purposes to determine the applicable law for an unrelated business purpose would be quite arbitrary. This is particularly true in light of the reality that some or all of the functions involved in the maintenance and servicing of a securities account are increasingly being undertaken from more than one office or outsourced to third parties in several locations.

Int-44 In modern global trading, the various activities involved in the maintenance of securities accounts are indeed often dispersed among offices in different countries. By way of illustration, assume an intermediary incorporated under the laws of the State of New York agreed with its client that the client's securities account is maintained in Tokyo because it is the place where the account was initially opened and where the first credit of securities to the account was effected. The intermediary sends all its account statements to the client from an office in Dublin. The client receives dividends administered and sent from an office in Hong Kong and obtains advice as to the ongoing status of the account from an office near the client's main office in Singapore. All the intermediary's operations relating to each of its securities accounts (including the entries effectuated) are backed-up and monitored by two separate computer systems run from its New Delhi and San Francisco offices respectively. Finally, the client regularly accesses the information relating to the relevant securities account from a laptop while travelling around the world.

Int-45 In such a situation, if the criterion for determining the applicable law were the location of the securities account or the location of the office where the securities account is maintained, no certainty could be achieved and such a test would invite litigation in which courts would be required to make fact-intensive inquiries. The risks and burdens presented to a potential collateral taker are readily apparent.

Int-46 Against this background, it became apparent in the course of the negotiations that the Convention had to move beyond the initial formulation of the PRIMA principle in order to provide the necessary *ex ante* legal certainty and predictability. It did this in two ways: (i) it abandoned the focus on attributing a "location" to an intermediary or a securities account and replaced it with an approach giving effect to an express agreement on governing law between an account holder and its intermediary and (ii) it added a Qualifying Office requirement. In doing so, it adhered to the agreed rejection of rules based on *lex rei sitae* or any look-through approach, retaining the notion of the relevant intermediary and focusing on the relationship between an account holder and the relevant intermediary with respect to a particular securities account.

5. The basis for the Convention's primary rule: an express choice of law agreement between the account holder and the relevant intermediary

Int-47 The Convention's primary conflict of laws rule (Art. 4(1)) is not based on any attempt to 'locate' a securities account, the office that maintains a securities account, the issuer, or the underlying securities. Rather, the Convention's primary rule is based on the relationship between an account holder and its intermediary: it allows the parties to an account agreement to expressly agree on the law governing all the issues falling within the scope of the Convention. This choice may be expressed in either of two ways: If the parties expressly agree on a law governing their account agreement (general governing law clause), that law also governs all the issues falling within the scope of the Convention; if, however, the account holder and its relevant intermediary expressly agree that the law of a particular State will govern all the issues falling within the scope of the Convention, that law governs all these issues (whether or not there is also a separate choice of law to govern the account agreement generally). The parties' choice of law, however, will be effective only if, at the time of the agreement on governing law, the relevant intermediary has an office ('Qualifying Office') in the selected State which, alone or with another office of the intermediary or a third party acting for the relevant intermediary, serves certain functions relating to securities accounts (though not necessarily the particular account in question), or is identified, by any specific means, as maintaining securities accounts in that State (though not necessarily the particular account in question).

Int-48 Article 4(1) reflects a policy decision to formulate a conflict of laws rule that reduces risk, promotes capital formation and investment, reflects existing and foreseeable market practice, is practical and efficient, and permits market participants to determine in advance which law governs all the issues falling within the scope of the Convention, providing *ex ante* legal certainty and predictability for the largest number of transactions. Under current market, legal and regulatory conditions, none of the alternatives considered and rejected would have provided the degree of *ex ante* legal certainty, simplicity, logic and clarity that the rule in Article 4(1) provides.

III. Key features of the Convention

Int-49 The Hague Securities Convention is a pure conflict of laws convention. It does not impose any changes on existing or future substantive law.

Int-50 The Convention's fundamental goal is to provide legal certainty and predictability as to the law governing issues that are of crucial practical importance for holdings and dispositions of securities held with an intermediary, in particular their effects against the intermediary and third parties. This purpose is achieved by the creation of a uniform conflict of laws regime (Arts. 4, 5 and 6) that displaces any national conflict rules in this matter and that provides the parties with the highest possible assurance as to which substantive law is applicable in their specific situation.

Int-51 The Convention applies only to issues relating to securities held with an intermediary, regardless of how the rights resulting from a credit of securities to a securities account are classified under applicable law or whether the account holder would be treated as the direct holder of the underlying securities or have the right to enforce the securities directly against the issuer under applicable law. Thus, until securities have entered the intermediated system by being credited to a securities account, the Convention does not apply to them. Matters relating to directly held securities (see para. Int-24) are outside the Convention's scope. The sole exception is that provided by Article 1(4) which extends the meaning of intermediary (Art. 1(4), in turn, is qualified by Art. 1(5); see comments in paras. 1-37 *et seq.*).

Int-52 The term securities (Art. 1(1)(a)) encompasses all instruments and assets that are financial in nature (other than cash) and which may be credited to a securities account. It does not matter whether the securities are in bearer or registered form and whether they are represented by a certificate or dematerialised. The term encompasses all types of debt and equity securities.

Int-53 The Convention applies whenever a situation involving securities held with an intermediary relates in any way to more than one State (Art. 3). This provision is deliberately very broad, and the Convention's sphere of application is not limited by traditional notions of internationality.

Int-54 The Convention law (*i.e.*, the law determined by the Convention's conflict of laws regime in Arts. 4, 5 and 6) applies to all the issues enumerated in the exhaustive but very broad and intentionally very generally worded list in Article 2(1). An issue not specified in Article 2(1) does not fall within the Convention's scope and, therefore, is not governed by the substantive law determined by application of the Convention rules contained in Articles 4 or 5.

Int-55 The Article 2(1) issues include in particular the legal nature and effects against the intermediary and third parties of rights resulting from a credit of securities to a securities account, as well as the law governing legal nature and effects against the intermediary and third parties of a disposition of securities. The Convention law governs the classification (if of any relevance under that substantive law) as contractual or not of the rights resulting from a credit of securities to a securities account (see comments in para. 2-30).

Int-56 The Convention also determines the law applicable to the perfection requirements of a disposition and whether an interest extinguishes or has priority over another person's interest; the law determined by the Convention will govern, inter alia, the priority between (i) a person who acquired an interest in securities in good faith, for value and without notice of an adverse claim (a so-called "bona fide purchaser" (BFP) or "protected purchaser") and (ii) an adverse claimant.

Int-57 Further, the Convention law also governs whether an intermediary has any duties to a person other than the account holder who asserts, in competition with the account holder or another person, an interest in securities held with that intermediary. This includes the question whether so-called upper-tier attachments are permissible (*i.e.*, attachments of an account holder's interest at a level above that of its own intermediary).

Int-58 The requirements for the realisation of an interest in securities held with an intermediary are also governed by the Convention law. Finally, the Convention law governs whether a disposition of securities held with an intermediary extends to entitlements to dividends, income, or other distributions, or to redemption, sale or other proceeds.

Int-59 The Convention does not determine the law applicable to matters not falling within Article 2(1), for example, such purely contractual or other purely personal rights between an account holder and its intermediary *inter se* as the content and frequency of account statements, the intermediary's standard of care in maintaining securities accounts, risk of loss, deadlines in giving instructions, and the like. Similarly, the Convention law does not extend to matters related to a disposition of securities not falling within Article 2(1) such as the number and type of securities to be disposed or the price for the securities. Finally, the

Convention does not determine the law applicable to the rights and duties of an issuer of securities or of an issuer's registrar or transfer agent. This exclusion encompasses the duties of an issuer with respect to all corporate actions, including voting rights, dividend rights and registration rights, and the rights of an issuer to define the steps for achieving good discharge of a note, bond or other debt security. Finally, the Convention has no impact on regulatory schemes relating to the issue or trading of securities, regulatory requirements placed on intermediaries, or enforcement actions taken by regulators.

Int-60 Under the Convention's primary rule (Art. 4), the applicable law is determined on the basis of an express governing law agreement between the account holder and the relevant intermediary, if that agreement is articulated in either of two ways: If an account holder and its relevant intermediary expressly agree that the law of a particular State[19] will govern their account agreement, that law also governs all the Article 2(1) issues; if, however, the account holder and its relevant intermediary expressly agree that the law of a particular State will govern all the Article 2(1) issues, that law governs all these issues, whether or not there is also a separate choice of a law to govern the account agreement. The parties may expressly agree to have the law of one State govern all the Article 2(1) issues and that of a different State govern the account agreement. The Convention law, however, applies to *all* of the Article 2(1) issues. It is not possible that some of these issues will be governed by one law while others are governed by a different law, nor is it open to the parties to select the law to govern only some of the Article 2(1) issues.[20]

Int-61 The law chosen by the parties to the account agreement applies only if the relevant intermediary has, at the time of the agreement on governing law, an office ('Qualifying Office') in the State whose law is selected which, alone or with another office or a third party (which does not have to be in the selected State), serves certain functions relating to the maintenance of securities accounts (though not necessarily the particular account in question), or is identified, by any specific means, as maintaining securities accounts in that State (though not necessarily the particular account in question).

Int-62 If the primary rule does not apply, either because the parties have expressly selected an applicable law but the Qualifying Office requirement is not satisfied or because they have failed to make any express selection in their agreement or have never concluded an account agreement at all (see para. 5-7), the Convention provides three fall-back rules which operate as a cascade (Art. 5). The determinative elements used in the fall-back rules are (in sequence): the law of the place of the office that a written account agreement expressly and unambiguously identifies as the office through which the relevant intermediary entered into the account agreement (Art. 5(1)); the law of the place of incorporation or organisation of the relevant intermediary (Art. 5(2)); and the law of the (principal) place of business of the relevant intermediary (Art. 5(3)).

Int-63 Articles 4 and 5 must be applied independently with respect to each securities account (*i.e.*, to each relationship between an account holder and its relevant intermediary).

19 For brevity, the expression "State" includes territorial unit of a Multi-unit State where applicable.

20 Thus, in this Report a reference to "the Article 2(1) issues" means *all* such issues.

Thus, the conflict of laws provision of the Convention will not assure, or even typically produce, the result that one single law will govern all the issues specified in Article 2(1) with respect to all securities accounts involved in dispositions of securities involving a chain of intermediaries (see paras. 4-43 *et seq.*).

Int-64 Article 6 sets forth a list of factors that must be disregarded when determining the applicable law under the Convention.

Int-65 Article 7 determines the impact, if any, under the Convention of an amendment to an account agreement if the consequence of the amendment is that the Convention law changes from the law of one State as determined under either Article 4(1) or Article 5, to the law of a different State as determined under Article 4(1). The principle is that the "new law" governs all the Article 2(1) issues in respect of any interest in securities previously or subsequently credited to the securities account governed by the amended account agreement (Art. 7(3)), subject to exceptions with respect to specified issues which are designed to protect certain interests in securities that were acquired before the amendment by a person who did not consent to the amendment (see Art. 7(4), which is fully discussed in paras. 7-16 *et seq.*). It is expected that Article 7 will come into play only rarely since the kinds of amendments to which it would apply are likely to be infrequent.

Int-66 The Convention provides in Article 8(1) that despite the opening of an insolvency proceeding, the Convention law will continue to apply to all Article 2(1) issues as regards pre-insolvency events. An interest acquired prior to the opening of an insolvency proceeding, under the law applicable under the Convention, must be recognised in the insolvency proceeding. Thus, an insolvency court or administrator may not refuse to recognise the right or its perfected status merely because this right had not (also) been created or perfected in accordance with the *lex concursus*. According to Article 8(2), however, it is left to the *lex concursus* to determine the effects of those rights, that is, the extent to which these rights can be used in the insolvency proceeding. Thus, insolvency rules continue to apply and, for example, a pledge perfected during a 'suspect period' may be invalidated as 'preference' or 'fraudulent conveyance' and the pre-insolvency priority may be displaced by an insolvency rule governing the ranking of claims.

Int-67 The law determined by the Convention is the applicable law whether or not it is the law of a Contracting State (Art. 9). Furthermore, the Convention does not leave any room for renvoi in the traditional private international law sense: the applicable law determined by the Convention refers to substantive rules only, not to conflict of laws rules (Art. 10; see, however, paras Int-71 and Int-72).

Int-68 Article 11 carefully restricts the grounds for judicial refusal to apply the law determined to be applicable under either Article 4 or 5 of the Convention. Article 11(1) sets forth a traditional public policy (ordre public) exception to the application of the law otherwise applicable under the Convention. Article 11(2) provides that the Convention does not prevent the application of the forum's substantive rules that have to be applied regardless of the applicable law (*i.e.*, that are "mandatory" in the private international law sense). However, Article 11(3) makes clear that even the public policy exception and the internationally mandatory forum rule exception may not be used to impose requirements with respect to perfection or relating to priorities between competing interests, unless the law of the forum is the Convention law (see full discussion in para. 11-12).

Int-69 Article 12 sets forth several crucial interpretive and substantive provisions relating to the application of the Convention with regard to Multi-unit States. In particular, Article 12(1) explicates how the Convention's primary rule (Art. 4) operates in relation to Multi-unit States when the parties have designated the law of a particular territorial unit. Thus, if the parties have expressly agreed on the law of a particular territorial unit of a Multi-unit State (either as the law governing the account agreement or as another law governing the Art. 2(1) issues), the applicable law is the law of that specified territorial unit, if the relevant intermediary has a Qualifying Office anywhere in the Multi-unit State. A Multi-unit State may, however, impose a geographically more stringent condition and require that if, under Article 4, the applicable law is that of one of its territorial units, the relevant intermediary has a Qualifying Office within the designated territorial unit.

Int-70 Article 12(2)(a) provides that the term "law in force in a territorial unit of a Multi-unit State" includes both the law of that territorial unit and, to the extent applicable (under the law of either the unit or the State) in that unit, the law of the Multi-unit State.

Int-71 Article 12(2)(b) provides that if the conflict of law rules in force in a territorial unit of a Multi-unit State designate the law of another territorial unit of that State to govern perfection by public filing, recording or registration, the law of that other territorial unit governs that issue (internal renvoi).

Int-72 Under Article 12(3), a Multi-Unit State may file a declaration to the effect that if the applicable law, determined under Article 5, is that of the Multi-unit State or one of its territorial units, the internal conflict of laws rules in force in that Multi-unit State must be applied and it is those rules which will determine whether the substantive law of that Multi-unit State or of a particular territorial unit of that Multi-unit State shall apply (internal renvoi).

Int-73 Article 13 calls for uniform interpretation of the Convention. The importance of this principle is reinforced by the provision in Article 14 for Special Commission meetings to review the practical operation of the Convention.

Int-74 Article 15 provides that the Convention's conflict of laws rule (either Art. 4 or 5) applies to determine the law that governs the issue of priority of an interest acquired subsequent to the date the Convention entered into force for the forum vis-à-vis an interest acquired earlier.

Int-75 Article 16 helps market participants to avoid the costly and unnecessary exercise of amending pre-Convention account agreements or the opening of new securities accounts in order to enjoy the benefits of the Convention (note that Art. 16 allows for some declarations by Contracting States, see para. Int-76). This goal is achieved by confirming in Article 16(1) that the Convention applies to account agreements that were entered into, and securities accounts that were opened, before the Convention entered into force on the international plane, and by including in Article 16(3) and (4) interpretive rules that treat certain provisions of pre-Convention account agreements as having the effect of determining, for purposes of Article 4(1) of the Convention, the law applicable to all the Article 2(1) issues. Thus, if express words in a pre-Convention account agreement would, under the rules of the State whose law governs the contract, have the effect of determining the law governing any of the Article 2(1) issues, that law will govern all these issues, but only if the intermediary met the Qualifying Office test at the time of the agreement on governing law. Similarly, if the parties have agreed, pre-Convention, that the securities account is maintained in a particular State, the law of that State will be the applicable law under the Convention, but only if the intermediary met the Qualifying Office test at the time of the agreement on governing law; in this particular situation, the agreement as to where the securities account is maintained may be express or implied from the terms of the contract considered as a whole or from the surrounding circumstances (subject to the prohibition in Art. 6).

Int-76 Under Article 16(2), a State may declare that its courts will not rely on the interpretive rules of Articles 16(3) and (4) in applying Article 4(1) to account agreements entered into during the period between the date the Convention enters into force on the international plane (Art. 19(1)) and the date the Convention enters into force for that State (the "Gap Period"). If a Contracting State makes such a declaration, the courts of that State will apply Articles 4, 5 and 6 to Gap Period account agreements without any interpretive assistance from paragraphs 3 or 4. Under Article 16(3), a State may make a declaration that the courts of that State will not apply the interpretive rule of Article 16(3) if the parties to the account agreement have expressly agreed that the securities account is maintained in a different State than the State whose law would otherwise be applicable by reason of the interpretive rule of Article 16(3).

Int-77 Articles 17 to 24 contain the final clauses, including a provision (Art. 18) that enables a Regional Economic Integration Organisation (REIO) constituted by sovereign States to sign, accept, approve or accede to the Convention, but only to the extent it has exclusive competence over matters covered by the Convention.

Int-78 Finally, it is important to stress the following with respect to the *travaux préparatoires*. Although there are occasional references in this Report, for historical purposes, to the Preliminary Documents or the Minutes, it must be noted that those materials should be used with caution. It is the function of this Report to explain the meaning of the Convention as actually agreed, and it will be only rarely that the *travaux préparatoires* might illuminate that meaning. Many crucial substantive as well as linguistic decisions were made very late in the process (often in the Diplomatic Session). Thus, earlier versions will shed little light on the meaning or intent of the final text. Further, the Preliminary Documents often represented merely proposals for discussion purposes (often proposals later rejected) or reactions to proposals, rather than tentatively agreed solutions. Finally, it must be recognised that the Minutes do not purport to be a verbatim transcript and are often incomplete. It is the final text and this Report which were reviewed and accepted by the Member States.

PART II:
Commentary on the Articles of the Convention

CHAPTER I DEFINITIONS AND SCOPE OF APPLICATION

Article 1 *Definitions and interpretation*

1. In this Convention –

a) "securities" means any shares, bonds or other financial instruments or financial assets (other than cash), or any interest therein;

b) "securities account" means an account maintained by an intermediary to which securities may be credited or debited;

c) "intermediary" means a person that in the course of a business or other regular activity maintains securities accounts for others or both for others and for its own account and is acting in that capacity;

d) "account holder" means a person in whose name an intermediary maintains a securities account;

e) "account agreement" means, in relation to a securities account, the agreement with the relevant intermediary governing that securities account;

f) "securities held with an intermediary" means the rights of an account holder resulting from a credit of securities to a securities account;

g) "relevant intermediary" means the intermediary that maintains the securities account for the account holder;

h) "disposition" means any transfer of title whether outright or by way of security and any grant of a security interest, whether possessory or non-possessory;

i) "perfection" means completion of any steps necessary to render a disposition effective against persons who are not parties to that disposition;

j) "office" means, in relation to an intermediary, a place of business at which any of the activities of the intermediary are carried on, excluding a place of business which is intended to be merely temporary and a place of business of any person other than the intermediary;

k) "insolvency proceeding" means a collective judicial or administrative proceeding, including an interim proceeding, in which the assets and affairs of the debtor are subject to control or supervision by a court or other competent authority for the purpose of reorganisation or liquidation;

l) "insolvency administrator" means a person authorised to administer a reorganisation or liquidation, including one authorised on an interim basis, and includes a debtor in possession if permitted by the applicable insolvency law;

m) "Multi-unit State" means a State within which two or more territorial units of that State, or both the State and one or more of its territorial units, have their own rules of law in respect of any of the issues specified in Article 2(1);

n) "writing" and "written" mean a record of information (including information communicated by teletransmission) which is in tangible or other form and is capable of being reproduced in tangible form on a subsequent occasion.

2. References in this Convention to a disposition of securities held with an intermediary include –

a) a disposition of a securities account;

b) a disposition in favour of the account holder's intermediary;

c) a lien by operation of law in favour of the account holder's intermediary in respect of any claim arising in connection with the maintenance and operation of a securities account.

3. A person shall not be considered an intermediary for the purposes of this Convention merely because –

a) it acts as registrar or transfer agent for an issuer of securities; or

b) it records in its own books details of securities credited to securities accounts maintained by an intermediary in the names of other persons for whom it acts as manager or agent or otherwise in a purely administrative capacity.

4. Subject to paragraph (5), a person shall be regarded as an intermediary for the purposes of this Convention in relation to securities which are credited to securities accounts which it maintains in the capacity of a central securities depository or which are otherwise transferable by book entry across securities accounts which it maintains.

5. In relation to securities which are credited to securities accounts maintained by a person in the capacity of operator of a system for the holding and transfer of such securities on records of the issuer or other records which constitute the primary record of entitlement to them as against the issuer, the Contracting State under whose law those securities are constituted may, at any time, make a declaration that the person which operates that system shall not be an intermediary for the purposes of this Convention.

I. Definitions – Article 1(1)

a) "securities"

1-1 The definition of "securities" is intentionally very broad, avoiding a list of included or excluded instruments which might become out of date because of developments in market practice. The term extends to all instruments and assets that are financial in nature (other than cash), whether they are in bearer or registered form and whether they are represented by a certificate or dematerialised. It thus encompasses all types of debt and equity security and is limited only by the fact that the instruments or assets in question must be financial instruments or assets and must in addition be of a kind capable of being credited to a securities account with an intermediary.

1-2 Financial instruments or assets typically embody a tradable entitlement to money (with or without other rights, such as a right to vote and/or membership rights) and, where not intended to be held solely by a single investor, are issued on terms standard for each unit of the issue with a view to being held, directly or through intermediaries, as a medium for investment. However, it is not necessary that financial instruments or assets credited to a securities account have any of these characteristics in order to constitute securities for purpose of the Convention. The definition is intentionally drawn in very broad terms so as to accommodate changes in market practice (see further comments in para. 1-3). Thus the Convention avoids setting out any "laundry list" of included and excluded instruments and relies on the phrase "held with an intermediary" as the primary limiting factor. Thus no court or legal adviser need be long delayed by the effort to determine whether an asset is a security; if an asset is of a kind credited to a securities account and is financial in nature it is a security within the meaning of the Convention. The manner of issue of an instrument is immaterial. The instrument may be issued directly on the market or by placement with an underwriter, a restricted group of investors or a sole investor. It may be issued in the form of a single global (or jumbo) certificate deposited with an international or national central securities depository in which investors can acquire an interest through an account with the depository, whether the global certificate is permanent or is temporary or semi-permanent, or permits the account holder in given conditions to exchange its interest for definitive certificates. Alternatively the instrument may (as in the case of certificates of deposit) be available by purchase from a broker, bank or other financial institution. Clearly falling within the definition are issued shares, stock and bonds (whether or not listed on a stock exchange or other official list), units in collective investment schemes, exchange-traded financial futures and options, credit derivatives, short-term promissory notes (commercial paper) issued on a money market, warrants, certificates of deposit, bankers acceptances, securities issued under Islamic law and American Depositary Receipts.

1-3 Other instruments, if not already considered to be securities, may become so as the result of evolving market usage. Thus the question whether a given instrument or other asset falls within the definition of securities depends not on the label attached to the instrument or asset, still less on national regulatory law, but rather on whether it is a type of financial asset capable of being credited to a securities account. The test of what constitutes securities is thus a fluid one which is to be determined by reference to the practice and perception of the relevant market at the material time.

1-4 The phrase "or any interest therein" makes it clear that the provisions of the Convention relating to the holding and disposition of securities are not confined to full ownership and co-ownership (including co-ownership which confers against the relevant intermediary property rights in relation to underlying securities) but extend to lesser interests, for example, possessory and non-possessory security interests. See also the comments in paragraph 1-19 on the meaning of "disposition", as well as the commentary on Article 2 (in particular the comments in paras. 2-30 and 2-31 on Art. 2(2), and the comments in para. 2-11 on Art. 2(1)(a)). Moreover, in light of the intermediated holding system, which may involve a chain of intermediaries between the account holder and the issuer, the phrase also refers to the interest which the account holder's intermediary (or any other intermediary in the chain) has in securities held with its intermediary.

1-5 The definition of securities expressly excludes "cash". The term "cash" is not confined to physical money (which in this context is of minimal significance) but is used in a broad sense to encompass bank deposits. Thus "securities" does not extend to cash paid to the intermediary towards the cost of acquiring securities on behalf of the account holder or providing margin cover, still less to ordinary deposit accounts. Questions as to an account holder's rights resulting from the credit of cash to an account held with the intermediary are governed not by the Convention law but by the applicable law as determined by the other conflict rules of the forum, and this is so whether the cash is credited to the securities account or to a separate cash account maintained by the intermediary. An entirely separate question is whether a disposition of securities held with an intermediary carries with it the rights to dividends, income or other distributions, or to redemption, sale or other proceeds received in respect of securities credited to a securities account. That question is governed by the Convention law under Article 2(1)(g) (see para. 2-29).

b) "securities account"

1-6 A securities account is an account maintained by an intermediary to which securities may be credited or debited. The account may exist at any level from the lowest level account holder to a central securities depository (CSD) or an international central securities depository (ICSD). In the context of a statutory operator of a securities transfer and settlement system, such as the United Kingdom CREST system and the Australian CHESS system (see para. 1-37), the term "securities account" is to be interpreted broadly to include any form of record of entitlements and transfers, whether or not denominated as an account by the operator. For the purpose of the definition it is not necessary for there to be any securities in the account at any particular time; all that is required is that securities "may" be credited or debited to the account. The account may be opened in anticipation of securities being credited to it, while securities that have been credited to the account may be disposed of, but until the account is closed it reflects a continuing contractual relationship between the account holder and the intermediary in relation to existing or after-acquired securities, a relationship which is the essence of the status of intermediary (see comments in para. 1-13).

1-7 The securities account will typically be governed by a written or oral account agreement. Nevertheless, Article 5(2) envisages the possibility of an account without an account agreement, as in the case of a Nordic CSD which is an intermediary by virtue of Article 1(4) and which maintains accounts governed not by an account agreement but by law and the rules of the CSD (see comments in para. 5-7).

1-8 What steps are required to constitute a credit to an account and when a credit is deemed to occur vary from one jurisdiction to another. In some jurisdictions it is necessary for an entry to be made; in others it suffices that a decision has been taken by the intermediary to credit the account and/or that the necessary processing has been initiated but not completed. These are matters which are to be determined by the Convention law. See Example 2-1.

1-9 As to the relevant securities account where the account holder makes a disposition in favour of its own intermediary, see the comments in paras. 4-41 and 4-42 relating to Article 4(3).

c) "intermediary"

1-10 The term "intermediary" is not limited to a person that is licensed, otherwise approved, registered or denominated as an intermediary. The term person is not defined in the Convention. It includes a natural person, or group of natural persons, a legal person including a corporation, a limited liability company, a partnership (whether or not recognized as a person separate from its partners), a trust, and a government, government agency or instrumentality, or political subdivision. Any person, or group of persons, whether natural or legal, who in the course of a business or other regular activity maintains securities accounts for others or both for others and for its own account is an intermediary when acting in that capacity. The phrase "or both for others and for its own account," coupled with Article 4(3), makes it clear that a person acts in the capacity of intermediary even when it is the party in whose favour the account holder makes a disposition, as where the account holder pledges securities, or transfers title to them, as security for an advance or by way of margin cover. Examples of intermediaries include CSDs and ICSDs, clearing corporations, securities firms, banks and central banks that engage in a business or regular activity of maintaining securities accounts for others, or for others and themselves, and are acting in that capacity.

1-11 A person is an intermediary only if it maintains securities accounts for others or both for others and for its own account and interests in securities are transferred across its books. That is apparent from several provisions of the Convention: the definition of "account agreement" in Article 1(1)(e); the exclusion of those acting as manager or agent under Article 1(3)(b); and the reference in Article 2(1)(a) to effects against the intermediary of rights resulting from a credit of securities to a securities account. See further the comments in paragraphs Int-17 *et seq.*, paragraph 1-16, as well as in paragraphs 1-32 *et seq.* and 4-43 *et seq.*

1-12 In the modern securities holding system, any intermediary below the first-tier intermediary (*i.e.*, the intermediary which itself or through a nominee holds securities directly from the issuer) will usually act in two different capacities. As a person holding securities with a higher-tier intermediary it is an account holder; as one who maintains accounts for its own customers it is an intermediary. See further comments in paragraphs 1-33 to 1-35.

d) "account holder"

1-13 An account holder is a person in whose name a securities account is maintained. Whether the named account holder is serving as the agent or trustee for another person, whether the identity of the principal or beneficiary is disclosed, whether the principal's or beneficiary's name is also mentioned in the title of the securities account, and whether the agency or fiduciary capacity of the account holder is disclosed, are all irrelevant to (and in no way detract from) the account holder's status as such under the Convention. Moreover, the presence of any or all of these facts does not make the principal or beneficiary an account holder. An account holder may authorise another person to give instructions to the intermediary with respect to the account, but that authority (whether or not revocable) does not constitute the other person an account holder, even if that other person's name is mentioned in the records of the intermediary relating to that securities account or, indeed, is designated in the title of the account.

1-14 As noted above, an intermediary may itself be an account holder with its own intermediary.

e) "account agreement"

1-15 This is the agreement between the account holder and its intermediary governing their respective rights and duties in relation to securities that are or may become credited to a securities account maintained by that intermediary. The definition does not require that the account agreement fulfil any formal requirements. It may be oral or in writing or partly oral and partly in writing; it may incorporate in whole or in part rules or the procedures of the intermediary; and if in writing may consist of one or more documents. However, the Article 5(1) fall-back rule applies only to *written* account agreements. As to the relevant account agreement in relation to a disposition of securities by an account holder to its own intermediary, see Article 4(3) and the comments in paragraphs 4-41 *et seq.* There are some situations in which there is no account agreement relating to the account and the rights of the parties are prescribed by the law (see comments in paras. 1-7 and 5-7).

f) "securities held with an intermediary"

1-16 This is a key phrase of the Convention and refers to the bundle of rights resulting from the credit of securities to a securities account, whether those rights are proprietary, contractual, mixed or other in character under the applicable law (which under Art. 2(1)(a) is the law determined by the Convention itself). The Convention applies only to securities credited to a securities account; it does not apply to the rights held directly from the issuer by a person who is the registered holder of securities in records maintained by or for the issuer or who is in physical possession of certificates representing the securities. So until securities are first credited to a securities account, thereby entering the intermediated system, the Convention does not apply in relation to them. But once securities have been credited to a securities account, then all Article 2(1) issues in relation to (i) any rights resulting from such credit or to (ii) a disposition of any such rights are governed by the Convention law. This is so even if, in the case of a disposition, there has been no credit to the transferee's securities account. It suffices that such rights are derived, directly or indirectly, from a securities account, whether held by the transferee, the transferor or some other prior party. In some legal systems, for example those of France and Japan, rights resulting from the credit of securities to a securities account are nevertheless treated as a direct holding of the underlying securities and those maintaining securities accounts and standing between the account holder and the issuer are not themselves regarded as holders of the securities. Nevertheless these rights fall within the scope of the Convention because they result from the credit of securities to the securities account, not the entry in the issuer's register or the possession of certificates. This is made clear by Article 1(4), the effect of which is that, subject to Article 1(5), the person maintaining the account is to be treated as an intermediary even if under the relevant law the securities credited to the account are deemed to be held direct from the issuer. See further comments in paragraphs Int-17 *et seq.*, 1-32 *et seq.* and the comments in paragraphs 4-43 *et seq.*

g) "relevant intermediary"

1-17 This means the intermediary that maintains the particular securities account for the particular account holder. The term is used to distinguish the account holder's intermediary from all other intermediaries. In the modern securities holding system, tiers of intermediaries often stand between the account holder and the issuer of the underlying securities and between the counterparties to a disposition of securities held with an intermediary. For any securities account there is only one relevant intermediary, namely the intermediary who maintains that account. Thus, the Convention requires the applicable law to be determined separately for each account. The significance of this lies in the fact that the Article 2(1) issues arising in relation to any securities account will be governed solely by the law applicable in relation to that account under Article 4 or Article 5. They will not, for example, be governed by the law applicable to the Article 2(1) issues arising in relation to accounts maintained by a higher-tier intermediary – even in a Contracting State whose substantive law allows the account holder to look through its own intermediary and assert rights against a higher-tier intermediary. See further comments on Article 2(1)(e) in paragraphs 2-24 *et seq.*

1-18 As to the relevant intermediary in relation to dispositions by the account holder in favour of its own intermediary, see Article 4(3) and the comments in paragraphs 4-41 and 4-42.

h) "disposition"

1-19 The term "disposition" covers any transfer of title, whether outright or by way of security, and any other form of security interest. In relation to outright transfers of title, "disposition" includes sale and repurchase (repo), purchase and resale (reverse repo), transfers under sell/buy-back or buy/sell-back arrangements and stock loans. In relation to security transactions it covers both possessory and non-possessory security. However, it must be borne in mind that the Convention does not apply to dispositions of securities held directly by the grantor of the security interest (see para. Int-24). So a pledge of bearer securities by physical delivery of the relevant certificates by the pledgor/holder to the pledgee is not a disposition to which the Convention applies. Accordingly the reference to possessory security interests in the definition of "disposition" is directed to those legal systems (particularly civil law systems) which have a concept of delivery of possession of intangibles, so that a possessory security interest in securities may be created by book-entry. Article 1(2) further clarifies the meaning of the term "disposition" (see comments in para. 1-30).

i) "perfection"

1-20 In many legal systems a validly created security interest or other disposition, though effective as between the parties by virtue of their agreement, is not effective against a third party acquiring an interest in the subject-matter unless some further step has been taken. This step is conveniently termed perfection, which in Article 1(1)(i) is defined as the completion of any steps necessary to render a disposition effective against persons who are not parties to that disposition, including an insolvency administrator and general creditors in the insolvency proceedings. For example, for a consensual security interest to have any effects beyond the secured party and the debtor/pledgor, the applicable law may require that the secured party take control over a securities account, have the securities transferred to an account in the secured party's name, make a public filing, or take some other step. For certain dispositions, the applicable law may not require the parties to take any particular steps beyond their agreement in order to make the dispositions effective against third parties. Thus the applicable law may deem an intermediary to have a perfected security interest over a securities account that is pledged or transferred to it with no further step beyond the agreement itself being required for effectiveness against third parties. Conversely, in some systems such a requirement is not merely a perfection requirement but a constituent element of the disposition itself without which a purported disposition has no effect.

1-21 Perfection does not mean that the perfected disposition must have priority over and in all events be invulnerable to the claims of all third parties. Even if a security interest or other disposition is perfected, it is a question of *priority* (not of perfection) under the applicable law whether the perfected interest has priority over particular competing interests, such as other perfected interests, certain types of unperfected interest, purchasers who acquire a competing interest in good faith without notice of an adverse claim and creditors such as tax authorities who may enjoy statutory priorities. Likewise, it is a priority question, not a perfection matter, whether a security interest perfected by one method (*e.g.*, taking control over a pledged securities account) has priority over a security interest perfected by another method (*e.g.*, filing a public notice relating to the security interest).

j) "office"

1-22 The meaning of "office" is relevant to a number of provisions of the Convention. Its primary significance lies in the fact that where, under Article 4(1), an account holder and its intermediary expressly agree on a choice of the law of a particular State to govern the Article 2(1) issues, that agreement is effective for Article 4(1) purposes only if, at the time of that agreement (*i.e.*, the agreement on governing law), the relevant intermediary has an office in that State which alone or together with other offices of the relevant intermediary or with other persons acting for the relevant intermediary anywhere is engaged in a business or other regular activity of maintaining securities accounts or otherwise satisfies any of the criteria to be a "Qualifying Office" (see further comments in paras. 4-21 *et seq.*). This requirement is designed to prevent the parties from choosing the law of a State which has none of the connections specified in the second sentence of Article 4(1) with the intermediary's securities business. "Office" is also used in the first fall-back rule (in Art. 5(1)) in the requirement that a written account agreement expressly and unambiguously identify the office through which the intermediary entered into the account agreement (which office must be a Qualifying Office); and it features again in Article 12(4), which deals with offices in Multi-unit States and refers back to Article 4(1).

1-23 The definition does not specify any particular facilities or features required for the existence of an office. So "place of business" is to be interpreted in accordance with its ordinary meaning and is sufficiently flexible to include, for example, a ship on which a mobile branch regularly conducts its business and to take account of technological developments, such as the use of the Internet to conduct activities relating to securities and the maintenance of securities accounts. Already there are many "virtual banks" which communicate with their account holders solely by electronic means, so that account agreements are not concluded at any place where the bank has a physical presence. This does not occasion any problem as there is nothing in the Convention that requires a reference to the place where the account agreement is in fact concluded, though it is open to the parties to designate that place expressly under the first fall-back rule in Article 5(1). Nor does the fact that a place of business is an office for the purposes of the Convention imply that it is an authorised office for regulatory purposes.

1-24 The exclusion of a place of business which is intended by the intermediary to be merely temporary is designed to prevent evasion of the Qualifying Office requirement. The use of the word "intended" makes it clear that the issue is to be judged by reference to the intermediary's intentions at the time when the office is established and not with the actual duration of the office and its activities, which would not necessarily be known at the time of the agreement. So an office which, at the time the securities account is opened, is not intended by the intermediary to be purely temporary but in fact closes after a short time because it fails to achieve the success expected will still constitute an office within the definition. Again, an office does not cease to be a Qualifying Office for the purpose of Article 4(1) when the intention subsequently changes to one of temporary use. See further comments in paragraphs 4-27 *et seq.*

1-25 The definition refers to a place of business at which *any* of the activities of the intermediary are carried on. It is not necessary that those activities should include the maintenance of securities accounts or the functions described in Article 4(1)(a)(i) and (ii). But if they do not and if the alternative criterion in Article 4(1)(b) is not met, the office will not constitute a Qualifying Office for the purpose of Article 4(1) or the first fall-back rule in Article 5(1) but it may constitute a place of business for the purpose of the second and third fall-back rules in Article 5(2) and (3). The office must be that of the intermediary itself. The definition excludes a place of business of any person other than the intermediary, and under Article 6(d) no account may be taken of any such place in determining the applicable law. Thus the office of a subsidiary or other affiliate of the intermediary is not an office of the intermediary for the purposes of the Convention. This exclusion does not, however, override the clear provisions of the Qualifying Office condition in the second sentence of Article 4(1)(a), which expressly provides for the combining of the functions of an office of the relevant intermediary with the functions of not only another office of the relevant intermediary but also the functions, wherever performed, of any other person acting for the relevant intermediary (including a subsidiary or an affiliate) in order to satisfy the condition set forth in the second sentence of Article 4(1).

k) "insolvency proceeding"

1-26 The definition is identical to that used in Article 5 of the 2001 *United Nations Convention on the Assignment of Receivables in International Trade* (UN Convention). It is a broad definition intended to encompass the various types of proceedings under the insolvency law of any State, including interim proceedings. The definition applies irrespective of the nature of the debtor, the grounds for commencement of proceedings, the purpose (liquidation, reorganisation or other) and whether the proceedings are voluntary or involuntary. It is, however, confined to collective proceedings, that is, proceedings in which the insolvency administrator represents the general body of creditors, as opposed to proceedings available in some jurisdictions which are designed to provide an enforcement remedy for a particular creditor (*e.g.*, in England, an administrative receivership).

l) "insolvency administrator"

1-27 This definition is taken from those contained in Article 5 of the UN Convention (see para. 1-26) and Article 1(l) of the 2001 *Convention on International Interests in Mobile Equipment* (Cape Town Convention). The insolvency administrator can be a governmental authority or a private person, in either case vested with authority under the relevant insolvency law. It does not, however, include a person who assumes management powers on behalf of a particular secured creditor (see comments in para. 1-26) or by a group of creditors under an informal workout arrangement. The reference to a "debtor in possession" reflects the insolvency laws of some jurisdictions by which the conduct of the business of an insolvent debtor undergoing reorganisation is left in the hands of its management.

m) "Multi-unit State"

1-28 Some States comprise territorial units that have different systems of law. These may either be federal States, in which legislative power is distributed between the federal legislature and state or provincial legislatures, or non-federal States consisting of two or more territorial units with different legal systems. Article 12, together with Article 5, contains specific provisions for determination of the applicable law for Multi-unit States, the effect of which is that for certain purposes references to a State in the Convention are to be treated as references to the relevant territorial unit and certain references are to be treated as being to the State itself (*e.g.*, the Qualifying Office test in Art. 4(1)), while Articles 16 and 20 enable a Contracting State to make declarations as to territorial units.

n) "writing" and "written"

1-29 The broad definition is intended to keep pace with technological developments. It makes clear that electronic transmissions are considered writings if they can be reproduced in tangible form. The definition is based on that used in Article 1(nn) of the Cape Town Convention (see para. 1-27) but omits the phrase "and which indicates by reasonable means a person's approval of the record". Accordingly there is no requirement of authentication by a physical or electronic signature or otherwise. A written account agreement is required for the first fall-back rule in Article 5(1). Declarations by a Contracting State are also required to be notified to the Depositary in writing under Article 22(a).

II. Further definitional provisions

A. Disposition – Article 1(2)

1-30 Article 1(2) contains three discrete provisions to eliminate doubt and avoid error with respect to the scope of the term "disposition". First, Article 1(2)(a) clarifies that a "disposition of securities" includes disposition of an entire securities account as well as disposition of one, some, or all of the securities credited to a securities account. For example, it is possible under some legal systems to grant a security interest in some or all of the securities credited to an account at the time of the granting or in those and securities later credited to the account; either such disposition leaves the grantor as the account holder, with the power and right to engage in dealings with respect to the encumbered securities (except as may be limited by agreement). A grant of a security interest in the entire account is an alternative technique to accomplish the grant of a security interest which automatically covers all securities then or thereafter credited to the account; this techniques facilitates the efficient providing of collateral while leaving the collateral giver the flexibility to manage its investments. Article 2(1)(a) makes clear that the latter technique is also within the meaning of the term "disposition." An outright transfer of a securities account differs in that the account would no longer be maintained for the transferor, who would thereafter no longer have the ability to deal with the securities in the account; indeed, an outright transfer of a securities account would actually be implemented by the transfer of all the securities then credited to that account to an account maintained for the transferee (rather than the transferor's account itself being transferred).

1-31 Secondly, Article 1(2)(b) states that references to a "disposition" include a disposition in favour of the account holder's intermediary. It is common for intermediaries to require a pledge or title transfer in their favour of securities held for an account holder to secure extensions of credit to the account holder, whether for the acquisition of securities or otherwise. Thirdly, although the Convention generally applies only to consensual interests, Article 1(2)(c) provides an exception by including as a "disposition" a lien by operation of law in favour of the account holder's intermediary if the lien is in respect of any claim that arises in connection with the maintenance and operation of a securities account. "Operation of law" encompasses all liens not merely based on agreement. Thus, Article 1(2)(c) extends to statutory liens and judicially created or recognised liens, including clearing liens and banker's liens, provided that in each case they relate to the maintenance and operation of a securities account, as opposed, for example, to a bank deposit account. Intermediaries that operate clearance and settlement systems often extend credit to their customers in order to facilitate the efficient operation of those systems. Again, broker-dealers may advance funds for the acquisition of securities on behalf of the account holder or may intend to look to the proceeds of other securities and then experience delay in settlement by the purchaser. Applicable statutory or judge-made law may give such intermediaries a lien over securities held for their customers to the extent of any credit provided by the intermediary as well as for charges involved in the administration of the account, the handling of corporate actions, and the like. Article 1(2)(c) ensures that such liens are treated as dispositions for purposes of applying the Convention's conflict of laws rules. Where a non-consensual lien falls outside Article 1(2)(c) (*e.g.*, a tax lien) this means only that it is not a disposition for the purposes of the Convention. It may nevertheless be affected by the Convention, for example because the Convention law determines the outcome of a priority dispute between the holder of a non-consensual lien and the grantee of a consensual security interest. A registrar or transfer agent is to be distinguished from a statutory operator of a securities transfer and settlement system (see para. 1-36).

B. Intermediary – Article 1(3) - (5)

1. Introduction

1-32 The provisions in Article 1(3) – (5) are designed to clarify whether certain persons (including certain systems and their participants) should be regarded as intermediaries for the purposes of the Convention. Organisations expressly identified during the discussions leading to the Convention as those to which these provisions are particularly relevant were the local central securities depositories (CSDs) in Denmark, Finland, Norway and Sweden, the CREST system for the transfer of securities of issuers in the United Kingdom and Ireland and the French and Japanese systems (see comments in paras. 1-11, and 1-16).

2. Article 1(3)

1-33 Article 1(3) makes it clear that a person is not an intermediary merely because it acts as registrar or transfer agent for an issuer of securities (Art. 1(3)(a)), or records in its books details of securities credited to securities accounts maintained by an intermediary in the names of account holders for which the person acts as manager or agent or otherwise in a purely administrative capacity (Art. 1(3)(b)).

1-34 The functions of a registrar typically include maintenance of the securities register and the handling of corporate actions on behalf of the issuer. The registrar is thus essentially a record keeper for the issuer and is not an intermediary. Commonly the function of registrar is combined with that of transfer agent, whose role is to handle transfers on behalf of the issuer, recording the name of the transferee in the register in place of that of the transferor. Again, a transfer agent does not in that capacity act as an intermediary. Article 1(3)(a) is strictly unnecessary but was inserted for the avoidance of doubt. This is reinforced by Article 2(3)(c) (see comments in para. 2-34). A statutory operator of a securities transfer and settlement system, such as the United Kingdom CREST system and the Australian CHESS system (see para. 1-37), may be distinguished from a registrar or transfer agent.

1-35 Sub-paragraph (b) is addressed to the practice by which a bank agrees with its customer to manage the customer's investments by arranging for a securities account to be opened with a third party in the name of the customer, the bank itself maintaining a parallel record of the customer's holdings. This does not make the bank itself an intermediary, since it is not the party holding the securities for the customer, nor can transfers be effected across the bank's books, which merely record what is held for the customer in the records of the third-party intermediary. The position of the bank is thus to be contrasted with that of a CSD or other person who maintains securities accounts across which transfers may be effected (see Art. 1(4) and comments in para. 1-36). Article 1(3) does not, of course, preclude a person that acts as an agent, manager or otherwise in a purely administrative capacity with respect to one account from being an intermediary with respect to another account.

3. Article 1(4)

1-36 This provision, which operates subject to Article 1(5), confirms explicitly that a person is to be considered an intermediary for purposes of the Convention in relation to (i) securities that are credited to securities accounts that are maintained by that person in the capacity of a CSD, or (ii) securities that are otherwise transferable by book entry across securities accounts maintained by that person in a capacity other than as a CSD, such as a central bank that maintains accounts for book-entry government securities issued by its government or a person that maintains securities accounts under statutory arrangements for the dematerialised holding and transfer of securities. In contrast to the persons described in Article 1(3), those referred to in Article 1(4) are CSDs, central banks and others across whose books transfers may or must be effected and who are not merely acting for the issuer in the capacity of registrar or transfer agent. Thus a CSD or a central bank or other person that maintains records of an investor's holdings qualifies as an intermediary even if the credit of the securities to the securities account maintained by the CSD, central bank or other person establishes a direct relationship between the investor and the issuer and leads to rights which are enforceable against the issuer. This is of particular relevance to the Nordic countries, where domestic securities, though credited to accounts maintained by a CSD, are nevertheless enforceable against the issuer, and it was desired that the CSD be treated as an intermediary, even when not acting in that capacity, in order to benefit from the legal certainty provided by the Convention. The effect is that the Convention applies in relation to an account with the CSD even though the account holder's right resulting from the credit of the securities to the securities account maintained by the CSD are enforceable against the issuer. It is not necessary to rely on Article 1(4) where the CSD is acting as a true intermediary, for example because it holds foreign securities in an account in its own name with another CSD and does so on behalf of its own account holders. Even a person that is not a CSD is to be treated as an intermediary if transfers may be effected across its books, and this is so whether or not the person concerned features in the chain of title. This provision also confirms that a central bank that acts with respect to securities that are transferable by book entry across securities accounts maintained by the central bank is an intermediary with respect to such securities. In a number of legal systems those maintaining accounts through which transfers may be effected by book entry are not links in the chain of title, which runs direct from the issuer to the ultimate account holder. Such is the case, for example, under Japanese and French law. They are nevertheless intermediaries for the purposes of the Convention. See also the comments in paras. Int-17 *et seq.*, 1-16 and 4-43 *et seq.* (in particular para. 4-50). As stated in paragraph 1-6, in the context of statutory operator of a securities transfer and settlement system, the reference to securities accounts is to be broadly interpreted to include any form of record of holdings and transfers, whether or not denominated by the operator as an account, and would currently thus include, for example, sub-registers maintained by the Australian CHESS system, which in other respects is similar to the United Kingdom CREST system described below.

4. Article 1(5)

1-37 Article 1(5) was drafted with the United Kingdom CREST system in mind, but may also be relevant for similar systems. The CREST system is a system for the transfer of securities constituted under the laws of the United Kingdom, Ireland, Jersey, Guernsey and the Isle of Man by computerised book entries in accounts maintained by CRESTCo Limited, the operator of the system. As operator CRESTCo maintains records of holdings of investors direct from the issuer. A person acquiring securities through CREST is credited in its records. In the case of securities constituted under UK law, the CREST records constitute the register of holders (and therefore the primary record of entitlement). In other cases, details of transfers effected through the CREST system are then transmitted to the issuer, which maintains the register of holders; the issuer is required by the applicable legislation to update the register to reflect the transfer. Article 1(4) would cause a problem where, as in the case of CREST, securities issued by a company incorporated in one State (Ireland) are entirely handled by a body such as CREST having no office in that State and incorporated and operational in another State (the United Kingdom). Article 1(4) could lead to the application of English law to Irish securities. The problem would not be resolved by Article 1(3)(a) as CREST is not a transfer agent for the issuer but a free-standing statutory operator. To deal with such cases Article 1(5) enables a Contracting State to declare that the operator of a system for the holding and transfer of securities on records of the issuer, or other records which constitute the primary record of entitlement as against the issuer, is not to be treated as an intermediary for the purposes of the Convention. CREST, under both the variants described above, falls within this description and may therefore be the subject of a declaration under Article 1(5). Article 1(5) also enables a Contracting State to exclude from the definition of intermediary an operator which is a CSD when it is not acting in the capacity of intermediary.

1-38 The power to make a declaration under Article 1(5) is limited to the State under whose law the securities are constituted, because that is the law that would otherwise be encroached upon by the regular operation of the Convention. This declaration may be made at any time.

1-39 A Contracting State (on the meaning of the term "Contracting State", see para. 1-43) need not make a declaration under Article 1(5) as to all types of securities issued under its law but may confine its declaration to one or more specified categories of securities.

1-40 A declaration under Article 1(5) only affects the relationship in which the CSD or other person treated as an intermediary by Article 1(4) is involved. The CSD may be a true intermediary as regards foreign securities which it holds through another CSD.

III. Additional terms used in this Explanatory Report

1-41 Certain terms which do not feature in the Convention are used for convenience in this Report.

1-42 **"Article 2(1) issues":** the issues listed in Article 2(1) as those which are governed by the Convention law.

1-43 **"Contracting State":** The Convention uses this term in numerous provisions but with varying meanings. According to Article 2(1)(f) of the 1969 Vienna Convention on the Law of Treaties, the term "contracting State" means a State which has consented to be bound by a Convention, whether or not that Convention has entered into force; this is in contrast to the term "party", which under Article 2(1)(g) of the Vienna Convention refers to a State that has consented to be bound by a Convention and for which that Convention is in force.[21]

1-44 **"Convention law"**: the substantive law determined by application of the Convention rules contained in Articles 4 or 5.

1-45 "**Qualifying Office**": an office that satisfies the condition specified in Article 4(1), second sentence (see comments in paras. 4-21 *et seq.*). In the case of Article 4(1) (primary rule), whether this condition is satisfied is determined as of the time when the express agreement on governing law is made or, in case of an amendment to the account agreement in which the previous agreement on governing law is either modified or singled out and expressly reconfirmed, as of the time of such modification or reconfirmation (see comments in para. 4-27 *et seq.*). In the case of Article 5(1) (fall-back rule), whether this condition is satisfied is determined as at the time the written account agreement (containing the specified express and unambiguous statement) is first entered into (and not as at the time of the making of an express agreement on governing law); see comments in paras. 4-29 and 5-4).

---

21 With a view to avoiding confusion, this Report follows the terminology of the Securities Convention and only uses the term "Contracting State". It should be noted, however, that in Art. 1(5) the Convention uses the term "Contracting State" to mean both a Contracting State and a State Party (either may make the envisaged declaration). Equally, in Art. 9 the term covers both a Contracting State and a State Party (the Convention law applies whether or not it is the law of a Contracting State or a State Party). In Art. 15, the Convention uses the term "Contracting State" to mean State Party. Art. 16(2) uses the term "Contracting State" three times. In the first instance, it means State Party; in the latter two instances, it means both Contracting State and State Party (*i.e.*, the envisaged declaration may be made by a State when it has either status). Art. 16(3) uses the term "Contracting State" twice. In both instances, it means both Contracting State and State Party (*i.e.*, the envisaged declaration may be made by a State when it has either status). Under Art. 18, the term "Contracting State" means both a Contracting State and a State Party.

Article 2 ***Scope of the Convention and of the applicable law***

1. **This Convention determines the law applicable to the following issues in respect of securities held with an intermediary –**

   a) **the legal nature and effects against the intermediary and third parties of the rights resulting from a credit of securities to a securities account;**

   b) **the legal nature and effects against the intermediary and third parties of a disposition of securities held with an intermediary;**

   c) **the requirements, if any, for perfection of a disposition of securities held with an intermediary;**

   d) **whether a person's interest in securities held with an intermediary extinguishes or has priority over another person's interest;**

   e) **the duties, if any, of an intermediary to a person other than the account holder who asserts in competition with the account holder or another person an interest in securities held with that intermediary;**

   f) **the requirements, if any, for the realisation of an interest in securities held with an intermediary;**

   g) **whether a disposition of securities held with an intermediary extends to entitlements to dividends, income, or other distributions, or to redemption, sale or other proceeds.**

2. **This Convention determines the law applicable to the issues specified in paragraph (1) in relation to a disposition of or an interest in securities held with an intermediary even if the rights resulting from the credit of those securities to a securities account are determined in accordance with paragraph (1)(a) to be contractual in nature.**

3. **Subject to paragraph (2), this Convention does not determine the law applicable to –**

   a) **the rights and duties arising from the credit of securities to a securities account to the extent that such rights or duties are purely contractual or otherwise purely personal;**

   b) **the contractual or other personal rights and duties of parties to a disposition of securities held with an intermediary; or**

   c) **the rights and duties of an issuer of securities or of an issuer's registrar or transfer agent, whether in relation to the holder of the securities or any other person.**

I. Introduction

2-1 Article 2 defines the substantive scope of the Convention not in general terms but by a comprehensive list of specific issues to which the Convention law applies.[22] The Convention is a pure conflict of laws convention and does not affect or provide substantive law applicable to securities held with an intermediary. Thus, it has no effect on existing or future rules of substantive law, in particular related to the nature of an investor's interest in securities held with an intermediary or the requirements for creating or disposing of such an interest; nor does it take a position on substantive law issues as to whether an interest in securities held with an intermediary can for any purpose be traced through that intermediary to a higher-tier intermediary or to the issuer (see paras. Int-18 *et seq.*, and the further comments below).

2-2 Article 2(1) enumerates, in an exhaustive but very broad and intentionally very generally worded list, all the issues falling within the scope of the Convention. The breadth of Article 2(1) is underscored by Article 2(2), which provides that the Convention determines the law applicable to all the Article 2(1) issues in relation to both a disposition of and a static interest in securities even if the rights resulting from the credit of those securities to a securities account are contractual in nature (see comments in para. 2-30). All the issues mentioned in Article 2(1)(a)-(g) are governed by the applicable law determined under Article 4 or one of the fall-back rules provided in Article 5. The list in Article 2(1) is intended to be comprehensive and to include all issues that might have practical significance; the Convention does not extend to issues not covered by Article 2(1). There will seldom be a need to consider the wording of the list in any detail or the particular category of the list into which an issue falls (see para. 2-9). It is not possible, in respect of a particular securities account, for some issues in Article 2(1) to be governed by one law and others by a different law (see para. 4-10).

2-3 The Convention applies only if securities are held with an intermediary. It does not apply to the rights of a person as registered holder of securities in records maintained by or for the issuer or as possessor of certificates representing bearer securities if the securities are not credited to a securities account (see Example 2-1 following para. 2-16). Until securities are first credited to a securities account, thereby entering the intermediated system, the Convention has no application. But once securities have been credited to a securities account, then all Article 2(1) issues in relation to any rights resulting from a credit or a disposition of such securities are governed by the Convention law, whether the credit is in favour of the account holder or anyone else and whether or not the account holder also has rights directly against the issuer, though the Convention does not determine the law applicable to the rights and duties of the issuer (Examples 2-1, 2-2, 2-3 and 2-12).

2-4 At the beginning of the proceedings, the focus had been on proprietary rights in respect of securities held with an intermediary; indeed, earlier drafts expressly limited the Convention to such rights. However, this approach was significantly modified in the course

22 For brevity these issues are referred to as the Article 2(1) issues, while the Convention law means the substantive law determined by the Convention (see para. 1-44).

of the proceedings and in the final text, which resulted from an extensive discussion at the Diplomatic Conference as to the need to cover all the Article 2(1) issues however they might be classified by a particular legal system (which led to the abandonment of the proprietary/contractual terminology as the basis for defining the scope of the Convention) and as to the relationship between Article 2(1) and what are now Articles 2(2) and 2(3). The intention, as reflected in the clarification of Article 2(1) by Article 2(2) and the subordination of Article 2(3)(a) and (b) to Article 2(2), was to distinguish (i) those rights (whether proprietary, contractual, mixed or other) which relate to either the securities themselves and result from the credit of securities to a securities account or the disposition of securities held with an intermediary, from (ii) those rights, to the extent they are not included within the Article 2(1) issues, which arise solely from the contractual relationship between the account holder and its intermediary or the parties to a disposition *inter se*. While the former fall under Article 2(1) and thus are governed by the Convention law, the latter are governed by the conflict of laws rules of the forum other than those contained in the Convention.

2-5 There may occasionally be cases where the question arises whether or not an issue falls within the Article 2(1) list. This question is to be answered by reference to the language of Article 2(1) and (2) and not by using the Convention law itself to characterise the rights. It is only when the applicable substantive law has been determined by the Convention that the Convention law comes into play to determine and give effect to any characterisation required by that substantive law. Thus, it is not necessary to classify an account holder's rights relating to the securities themselves and resulting from a credit of securities to a securities account as proprietary or personal or otherwise in order to determine whether the Convention applies. The Convention applies to the rights relating to securities credited to a securities account, however the legal nature of these rights is classified in any legal system and whether or not the account holder has rights directly against the issuer.

2-6 The division between issues specified in Article 2(1) and those that are not is also reflected in Article 4(1), which allows the parties to an account agreement to choose a law to govern the Article 2(1) issues which is different from the law governing the account agreement generally, the latter being determined by the conflict of laws rules of the forum other than those contained in the Convention.

2-7 Article 2(3)(a), the wording of which is not as felicitous as it might be in its reference to rights resulting from a credit of securities to a securities account, is not intended, and should not be read, as a qualification of Article 2(1) but rather as dealing with those purely contractual or other purely personal rights that fall outside Article 2(1) altogether. Hence, Article 2(3)(a) is subordinated to Article 2(2), which clarifies that the Convention applies to the rights relating to all securities held with an intermediary even if the rights resulting from a credit of securities to a securities account are determined to be contractual in character (see comments in para. 2-30). By contrast, the effect of Article 2(3)(a) is to underscore that Article 2(1) does not extend to such matters as the content and frequency of account statements, the intermediary's standard of care in maintaining securities accounts, risk of loss, deadlines in giving instructions, and the like. Similarly, the effect of Article 2(3)(b), which is also subordinated to Article 2(2), is to underscore that Article 2(1) does not extend to such matters related to a disposition of securities as the number and type of securities to be disposed or the price for the securities (see Example 2-11).

2-8 Accordingly Article 2(1) should be given the broadest interpretation so as to include all rights resulting from the credit of securities to a securities account, no matter how the legal nature of those rights is classified by any legal system. Included in the Article 2(1) issues are whether a repurchase agreement or transfer of title arrangement may be recharacterised as a pledge (Example 2-7), the power of a collateral taker to re-use securities held with an intermediary (Example 2-8), perfection requirements, priority rules (Example 2-9), the duties of an intermediary faced with competing claims, including 'upper tier' attachments (Example 2-10), and the question whether a disposition of securities carries with it a right to dividends and other income and proceeds.

II. Article 2(1): issues governed by the applicable law determined by the Convention – the provisions in detail

A. The nature of the list of issues

2-9 Article 2(1) defines the domain governed by the Convention law by means of a comprehensive catalogue of practical issues that typically arise in transactions involving securities held with an intermediary. The primary transactions are a credit of securities to a securities account and a disposition of such securities (which involves both a debit and a credit to one or more securities accounts with one or more intermediaries). It is readily apparent from the wording of the list that in any given case more than one sub-paragraph may apply to the same fact situation, so that there is nothing to be gained from determining which particular sub-paragraph applies. It suffices that the issue falls within the list. For example, the granting of a security interest in securities held with an intermediary is a disposition within Article 2(1)(b), and, accordingly, the Convention determines the law applicable to the "effects against the intermediary and third parties" of the security interest. Yet to render a security interest effective against third parties, the security interest must satisfy the relevant perfection requirements, which are referred to in Article 2(1)(c). Under Article 2(1)(d), the Convention law would determine whether the collateral taker's security interest extinguishes or has priority over another person's interest (including the interest of the account holder, the relevant intermediary and any third party). This example demonstrates that the list should be looked at as a whole, and only if the issue cannot fit into any of the sub-paragraphs is it not covered by the Convention.

2-10 The list in Article 2(1) serves another function. In respect of a particular securities account, all the issues specified in sub-paragraphs (a)-(g) are governed by the same law, that is, the law determined by the Convention. Likewise, Articles 4 and 5 operate to determine the law for all the specified issues. Thus, the list in Article 2(1) ensures that all practical issues that may arise in relation to transactions involving securities held with an intermediary are covered by the Convention, so that not only will the coverage of the law determined by the Convention be universally recognised but also a single law will extend to all such issues. The more the parties can rely on the applicability of the Convention, the more the Convention's goal of providing *ex ante* certainty will be achieved. The language and structure of Article 2 are designed to discourage courts from engaging in efforts to hold matters to be outside the Convention's scope. If the issue concerns securities credited to a securities account, it will likely fall within the broad language of Article 2(1).

B. The content of the list

(a) Legal nature and effects of rights resulting from a credit of securities to a securities account

2-11 Different legal systems characterise the account holder's rights resulting from a credit of securities to a securities account in rather different ways. In some systems, the account holder's rights are characterised or denominated as a regular deposit, special deposit or some other form of property right traceable to individual securities. In other systems, the account holder's rights are characterised as an irregular deposit, general deposit or some other form of a purely personal (contractual) right against the intermediary to the delivery or transfer of a given type and number of securities. In still other systems, the account holder's rights are characterised or denominated as the interest of a beneficiary under a trust, a fiduciary interest, a *Gutschrift in Wertpapierrechnung*, co-property rights in a fungible, notional or book-entry pool of securities, security entitlements or some other bundle of property, contractual or other rights (see Examples 2-4 to 2-6).

2-12 The need for a clear conflict of laws rule to determine which law governs the legal nature and effects of an account holder's rights resulting from a credit of securities to a securities account relates to all legal systems, independently of how the right of an investor against the relevant intermediary in relation to any securities held through it may be characterised. Article 2(1) addresses this need by providing that the Convention determines the law governing the legal nature and effects of such rights against the intermediary and third parties. Accordingly, whether an account holder's rights resulting from the credit of securities to a securities account are characterised in one of the ways described in the preceding paragraph or in any other way, the Convention will determine the law governing the Article 2(1) issues, and this will be so even where under the Convention law an investor is treated as the direct owner of the securities.

2-13 Article 2(1)(a) also applies to a situation where, after the initial credit of securities to a securities account, there is no subsequent disposition of the securities (an example of such a 'static' situation is where an investor buys securities and wants to know what rights it has against the intermediary as a result of the credit of the securities to the investor's securities account).

2-14 The Convention is concerned solely with securities held with an intermediary. It does not cover questions arising in relation to directly held securities. The Convention also does not determine the law applicable to the rights and duties of an issuer of securities, whether the securities are directly held or held with an intermediary (Art. 2(3)(c); see para. 2-34).

2-15 Securities may be credited to a securities account because, for example, (a) the issuer, in the course of the initial distribution of the securities, deposited them with a CSD, either physically or by book-entry, with instructions to credit them to the securities account of the account holder; (b) the account holder originally held the securities in its own name and then deposited them with its intermediary; or (c) the intermediary has acquired or received securities for the customer's account or has transferred them to that account from the intermediary's own existing holding. Whether and when the credit has been legally effected is determined by the Convention law (see Example 2-2), though it is only in a small minority of cases that the question will arise. For example, under the Convention law, securities may be considered as credited to a customer's account (even if not in fact already credited) when they have been received by an intermediary (*i.e.*, credited to that intermediary's account maintained by its own intermediary) for the benefit of the customer, in effect a 'deemed credit'.

2-16 If the forum court determines that, under the Convention law, the securities in question have not yet been credited to any securities account so as to enter the intermediated system, then the Convention does not apply in relation to those securities and other conflict of laws rules of the forum will govern any issues relating to the securities. But once securities have, under the Convention law, been credited to a securities account, the Convention law governs the Article 2(1) issues both in relation to a static situation (see the comments in para. 2-13) and in relation to any subsequent disposition, whether or not that disposition has itself been reflected in a credit to another securities account (see Example 2-2). However, once a disposition has been reflected by a credit to another securities account, the rights resulting from that credit, as well as all other Article 2(1) issues, will be governed by the Convention law determined with respect to that other securities account, and if that other securities account is maintained by a different intermediary it is that intermediary which will be the relevant intermediary (see further comments in paras. 4-11 and 4-43 *et seq.*).

Example 2-1

> Investor I is the holder of certain registered securities and is in possession of certificates representing other securities which are bearer securities. I sells all the securities to C, who becomes registered in place of I as the holder of the registered securities and to whom I delivers the certificates representing the bearer securities. The Convention does not determine any of the issues arising out of the interests of I or C in the securities or the disposition of the securities because the securities were never credited to a securities account. The conflict of laws rules of the forum other than those contained in the Convention will determine the law applicable to all the issues arising out of an interest in or the disposition of those securities.

Example 2-2

> The facts are the same as in Example 2-1, except that, in accordance with C's instructions, I delivers the certificates to C's intermediary to be credited to C's securities account. Subsequently C instructs its intermediary to credit the securities to T's account with T's intermediary but because of a computer error there is some delay before these are credited to T's account with its own intermediary. The Convention law determines when the securities will be deemed to have been credited to C's securities account and what the effect of that credit is. If, under the Convention law, the securities are deemed not to have been credited to that securities account, the Convention has no further application because the securities are not yet held with an intermediary and, until they are, the law governing interests in them will be determined by the conflict rules of the forum other than those contained in the Convention. If under the Convention law, the securities are deemed to have been credited to C's securities account and thereby entered the

intermediated system, the Convention law applicable in relation to that account will govern the Article 2(1) issues arising not only in relation to C's interest in the securities but also in relation to the disposition to T. Once the securities have been credited to T's securities account, the rights resulting from that credit, and all other Article 2(1) issues, will be governed by the Convention law determined with respect to T's rather than C's securities account. See paragraphs 4-43 *et seq.* In no event, however, will the Convention determine the law applicable to the rights and duties of the issuer, because those matters are not included in Article 2(1).

Example 2-3

An investor causes securities to be credited to a securities account maintained for it by an intermediary organised under the laws of Korea. The Korean intermediary holds a corresponding position in a like number of securities of the same issue in a securities account maintained for the Korean intermediary by an intermediary organised under the laws of Japan. The underlying securities were issued by a company organised under Japanese law. Later, in a proceeding in Ruritania, where the Convention is in force, the issue is whether the Convention applies to determine the law applicable to the Article 2(1) issues in relation to the investor's rights resulting from the credit to its securities account. The plaintiff argues that the Convention does not apply because, under Japanese law, the investor has the right to enforce the securities directly against the issuer and is otherwise treated as the direct owner of the securities. This argument is without merit and must be rejected. The Convention determines the law that governs the Article 2(1) issues in respect of securities held with an intermediary, independent of whether the investor has a right to enforce the securities directly against the issuer or is otherwise treated as the direct owner of the securities. In no event, however, will the Convention determine the law applicable to the rights and duties of the issuer, because those matters are not included in Article 2(1).

Example 2-4

An investor has securities credited to a securities account maintained for it by an intermediary organised under New York law. Later, in a proceeding in Ruritania, where the Convention is in force, the issue is whether the Convention applies to determine the law applicable to the Article 2(1) issues in relation to the securities credited to the securities account. The plaintiff argues that (i) the Convention does not determine the law applicable to any of the Article 2(1) issues to the extent the legal nature of the investor's interest in the securities is "purely contractual or otherwise purely personal" in nature, (ii) New York law defines the legal nature of the rights arising from the credit of securities to a securities account as a sui generis bundle of rights called a security entitlement and (iii) some or all of those rights are purely contractual or otherwise purely personal in nature. These arguments are fallacious. It is not necessary to classify the rights in question as proprietary or purely personal or otherwise in order to determine whether the Convention applies. The Convention determines the law applicable to all the Article 2(1) issues in relation to the investor's security entitlement, independent of whether the security entitlement is classified as partly or purely proprietary, contractual, mixed or other under New York, Ruritanian or any other law. In contrast, the Convention does not determine the law applicable to purely contractual matters between an account holder and its intermediary *inter se* which fall outside the matters listed in Article 2(1) (see comments in paras. 2-4 to 2-7, 2-32 to 2-33, and Example 2-11).

Example 2-5

The facts are the same as in Example 2-4, except that the intermediary is organised under the laws of Germany. The plaintiff argues that the Convention does not determine the law applicable to any of the Article 2(1) issues because German law defines the legal nature of the rights resulting from the credit of securities to a securities account with the intermediary as a Gutschrift in Wertpapierrechnung and such an interest is purely contractual or otherwise purely personal in nature. Even if this were true the Convention would still apply.

Example 2-6

The facts are the same as in Example 2-4, except that the intermediary is organised under the laws of Italy. The plaintiff argues that the Convention does not determine the law applicable to any of the Article 2(1) issues because Italian law defines the legal nature of the rights resulting from the credit of securities to a securities account with the intermediary as an irregular deposit and such an interest is purely contractual or otherwise purely personal in nature. This argument is without merit. It is not necessary to classify the rights in question as proprietary or purely personal or otherwise in order to determine whether the Convention applies. The Convention determines the law applicable to all the Article 2(1) issues in relation to the investor's irregular deposit, independent of whether the irregular deposit is classified as partly or purely proprietary, contractual, mixed or other under Italian, Ruritanian or any other law.

2-17 The position is the same even as to rights to securities which the parties themselves agree are to be purely contractual.

(b) Legal nature and effects of a disposition

2-18 Under Article 2(1)(b), the Convention law determines the legal nature and effects against the intermediary and third parties of a disposition of securities. The term "disposition" is defined in Article 1(1)(h) and is further elaborated in Article 1(2) (see comments on these provisions). These provisions give effect to the Convention's functional approach of including all transfers by not restricting the meaning of "disposition" on the basis of either the economic function of the disposition or the legal category into which it might be placed. Therefore, in order to determine whether the Convention applies, it is not necessary a priori to assess whether a disposition is, for example, an outright sale or a transfer of title by way of security; rather, it is the applicable law determined by the Convention which governs the nature of the disposition (and its third party effects).

2-19 Accordingly the parties to a disposition may rely on the fact that the legal nature (and consequences) of that disposition under the applicable law as determined by the Convention is not going to be questioned or altered under a different law (at least not by a judge in a State where the Convention is in force). This is particularly important with regard to a transfer of title for collateral purposes or a sale and repurchase agreement ("repo"). While repos have (at least in some legal systems) the advantage of avoiding both difficulties relating to perfection rules applicable to pledges and restrictions on the reuse of the securities by collateral takers, parties to repo transactions traditionally have faced the risk that the repo might be recharacterised as a pledge by a court – because the court determines that either (i) the parties intended the transaction to have the legal nature of a pledge regardless of its form or (ii) the legal nature of the transaction in substance is a pledge regardless of the parties' intent – with the consequence that if the perfection requirements for a pledge have not been satisfied, the whole collateral arrangement might be void. If parties enter into a repo transaction (or use any other mechanism of a transfer of title by way of security), all judges in all Contracting States are bound to apply the Convention law to determine the legal nature of that disposition. The form, substance or the parties' intent as to the legal nature of the transaction has no bearing on determining which law governs the legal nature of the disposition but may be relevant in applying the Convention law. The Convention does not provide substantive law with respect to the legal nature of the disposition (and, thus, does not preclude recharacterisation under the applicable law), but the Convention does determine the applicable law governing the issue and, thus, provides predictability for transacting parties.

Example 2-7

> An investor organised under the laws of Japan and a bank organised under the laws of England enter into an agreement pursuant to which: (1) the investor sells to the bank all the securities credited to a securities account maintained in the name of the investor by an intermediary organised under the laws of Singapore, and (2) the investor agrees to repurchase from the bank equivalent securities after 30 days at a specified price (the bank agrees to sell under the same conditions). This agreement contains a governing law clause designating English law. The bank holds all its securities in a securities account maintained by the same intermediary. The intermediary expressly agrees with all its customers (including the investor and the bank) that their account agreements are governed by the laws of Singapore. Assume that Singapore law is the Convention law applicable to all the Article 2(1) issues. Assume also that the Convention is in force in the United Kingdom. In a proceeding in an English court, the nature and effects of the repo transaction are called into question. Under Article 2(l)(b), the law applicable under the Convention determines the legal nature of a disposition. Regardless of its view of the form, substance, or the parties' intent as to the legal nature of the disposition, the English court is bound to apply the law of Singapore to that question and is not free to reach a different result by recharacterising the repo under English law (despite the fact that English law is both the law governing the repo agreement and the law of the forum) or any other non-Singapore law.

2-20 As noted above, Article 2(1)(b) provides that the law determined by the Convention determines the "legal nature" and the "effects" of a disposition "against the intermediary and third parties". These terms encompass the important issue whether the rights acquired by a person from a disposition of securities include the right to reuse the securities with or without the consent of another person. (This issue is, for the same reason, likewise encompassed when presented in the context of the "legal nature" and "effects" of the rights resulting from a credit of securities to a securities account.) Of course, the foregoing is true whether the transaction involves an outright transfer or one made for security purposes, and whether the legal nature of the interest in the particular securities held with an intermediary is proprietary, contractual, mixed or other. Under virtually all legal systems, an attribute of ownership of an asset or claim is the right to transfer it to a third party or otherwise use the asset or claim for the owner's own purposes without any consent from the person from whom it was acquired. The law differs from State to State, however, over whether and to what extent the rights acquired under a disposition by way of security include the right to repledge, rehypothecate or otherwise reuse the securities. The Convention does not provide substantive law to resolve this issue, but it does determine the applicable law governing the issue.

Example 2-8

An investor has securities credited to a securities account maintained for it by an intermediary. The investor grants a security interest to the intermediary in all the securities credited to its securities account to secure a line of credit from the intermediary. The security agreement between the investor and the intermediary is silent on whether the intermediary may lend, rehypothecate or otherwise reuse the pledged securities credited to the investor's account for the intermediary's own benefit. The intermediary lends all the securities credited to the investor's securities account to a third party, without obtaining the investor's consent. The Convention law determines whether the intermediary's interest in the securities included the right to reuse them without the investor's consent.

(c) *Perfection requirements*

2-21 "Perfection" is defined by Article 1(l)(i) as "completion of any steps necessary to render a disposition effective against persons who are not parties to that disposition." The Convention law determines the requirements for the perfection of a disposition of securities held with an intermediary. Perfection requirements vary from State to State. Although they vary as to acceptable methods and in their effects, they usually play a role in determining priorities. Often they serve as anti-fraud devices, providing objective evidence of a transaction and a "date certain" and also serve to publicize the transaction. Common methods of perfection include public filing, recording, registration and the taking of "control" (which in some States is achieved without debiting the collateral provider's account but instead by the agreement of the intermediary to accept instructions concerning the collateral from the collateral taker without further consent from the collateral provider (the account holder) even while leaving the collateral provider with the right and power to dispose of the collateral).

2-22 The fulfilment of any perfection requirements imposed by the Convention law, though necessary to render the disposition effective against third parties, does not guarantee the priority of a perfected interest over a competing interest. That is a matter which under sub-paragraph (d) is also to be determined by the Convention law, which may give certain types of interest priority even over previously perfected dispositions.

(d) Priority issues

2-23 Under this sub-paragraph the Convention law determines which of two or more competing claims to an interest in securities is superior. The competition may be between two absolute interests, two limited interests (*e.g.*, security interests) or an absolute interest and a limited interest. The referral to the Convention law of the issues specified in sub-paragraph (d) is broad – it includes not only the simple issue of priority, but also the effects of such a decision – that is, whether the competing interests co-exist, with one being preferred over the other, or one takes free of the other altogether. Taking free from a competitor's interest is tantamount to extinguishing the other interest as between these two competitors, but does not necessarily mean the extinction of the competitor's interest as far as third parties are concerned. The Convention does not resolve these priority and effects issues; it simply provides the conflict of laws rule to determine which substantive law will govern these matters.

Example 2-9

> Same facts as Example 2-8, except that the issue is whether the third party's interest acquires priority over or takes free of the investor's interest in the securities lent by the intermediary. The Convention law determines those issues, regardless of whether the intermediary had the right to reuse the securities without the investor's consent.

(e) Duties of an intermediary against a person who asserts a competing interest

2-24 This paragraph is concerned with the duties of an intermediary faced with competing claims to an interest in securities held with that intermediary. Thus, the Convention law determines, for example, whether (a) an intermediary that honours a transfer instruction from the party who has priority under the Convention law, has fulfilled its duty; (b) the intermediary is protected if it honours an instruction from one claimant even if it is later found that another claimant has priority; and (c) the intermediary must ignore one claim and respect another.

2-25 The Convention law determines the intermediary's duties when faced with competing claims, including a competition between:

(1) the account holder and a person who asserts a competing interest in the securities held with the intermediary;

(2) the account holder and a person asserting a right to attach such securities;

(3) parties to successive dispositions by the account holder, each claiming priority for its interest in such securities.

2-26 Of critical importance to the safe, sound and efficient operation of the modern securities holding system, under Article 2(1)(e), the applicable law determined by the Convention governs whether so-called *upper-tier attachments* are permissible. A person claiming to have a competing interest in securities credited to a securities account might attempt to assert that competing interest, not only against the account holder's intermediary, but also against any one of the other intermediaries (*i.e.*, upper-tier intermediaries) standing between the account holder's intermediary and the issuer of the underlying securities. Under Article 2(1)(e), the Convention determines the applicable law that governs whether there can be an upper-tier attachment, that is, whether an adverse claimant can assert its interest against an intermediary in competition with the intermediary's account holder or another person (including an account holder of a lower-tier intermediary). The Convention does not determine whether upper-tier attachments are permissible; rather, it determines which substantive law resolves that question. In determining what law governs this issue under the Convention it is important to keep in mind that (i) several intermediaries will often stand between an account holder and the issuer of securities credited to the account holder's securities account, (ii) the Convention applies independently to each account holder and its own intermediary, (iii) when securities are transferred, the transferee's rights and all the other Article 2(1) issues, remain governed by the Convention law determined with respect to the transferor's securities account for so long as the securities are credited to that account, but once the securities are credited to the transferee's securities account, the transferee's rights, and all the other Article 2(1) issues, are governed by the Convention law determined with respect to the transferee's securities account (see Example 2-2 and the comments in paras. 4-11 and 4-43 *et seq.*), and (iv) a person that is the relevant intermediary for a particular account holder may be an account holder with an (upper-tier) intermediary.

Example 2-10

Investor, organised under the laws of Japan, holds an interest in securities credited to a securities account maintained for it by an intermediary organised under the laws of the Philippines. The intermediary holds a corresponding position in the securities in a securities account maintained for the Philippines intermediary by a bank organised under the laws of Germany. The German bank holds a corresponding position in the securities in a securities account maintained for the German bank by a central securities depository (CSD) organised under the laws of New York, whose nominee is the registered owner of the entire issue of securities represented by a global certificate. A judgment creditor of Investor attempts to execute its judgment against Investor by bringing legal process against the CSD seeking to attach Investor's alleged "indirect" interest in a portion of the securities credited to the securities account maintained for the German bank by the CSD in order to satisfy the creditor's judgment against Investor. In a proceeding in a court in a State where the Convention is in force, the question arises whether the CSD has any duty to turn over such portion of the securities registered in the name of the CSD's nominee to the judgment creditor, or whether the judgment creditor's exclusive remedy is to bring an action directly against Investor or Investor's own intermediary. Assume that the Convention law in relation to any interest in securities credited to the securities account maintained for German intermediary by the CSD is New York law and the Convention law in relation to any interest in securities credited to the securities account maintained for Philippine intermediary by the German intermediary is German law. The question whether the CSD has a duty to turn over the securities to the judgment creditor would be answered by New York law because that is the Convention law in relation to any interest in securities credited to the securities account maintained for the German intermediary by the CSD.

2-27 Though Article 2(1)(e) is primarily concerned with cases where the intermediary is faced with a claim by an attachment creditor of the account holder or with other competing claims, it also applies to the liability of an intermediary who is alleged to have acted in breach of duty to a claimant other than the account holder. In such a case the Convention law determines, among other things, whether the intermediary did in fact owe a duty to that claimant, whether it breached that duty and, if so, what are the remedies of that claimant.

(f) Realisation

2-28 Under Article 2(1)(f), the requirements for the realisation of an interest in securities held with an intermediary are governed by the Convention law. For example, if, upon the default of a collateral provider, a collateral taker wishes to sell the collateral given to it by the collateral provider, the Convention law will determine whether it can do so and what conditions apply to the exercise of that power. Such conditions might under the applicable law include, for example, court approval of the sale and/or a sale by public auction rather than private agreement.

(g) Entitlements to dividends, income, or other distributions, or to redemption, sale or other proceeds

2-29 Article 2(1)(g) does not expand the meaning of the term "securities held with an intermediary" to include cash held with an intermediary. Nevertheless, the Convention law determines whether a disposition of securities held with an intermediary extends to any entitlements to dividends, income, or other distributions, or to redemption, sale or other proceeds. For example, if an investor granted a security interest in 1000 shares credited to a securities account, the Convention law will determine whether the security interest extends to entitlements to dividends paid by the issuer in respect of those shares. The items covered by Article 2(1)(g) may take a variety of forms, including dividends, interest, bonus issues, rights issues, proceeds of redeemed debt securities, proceeds of sales by the account holder, and conversion of securities into other securities.

III. Article 2(2): law applicable to a disposition of or an interest in securities when the account holder's interest is a contractual claim against its intermediary

2-30 Article 2(2) confirms explicitly that, notwithstanding Articles 2(3)(a) or (b), the Convention determines the law applicable to the Article 2(1) issues in relation to a disposition of or an interest in securities held with an intermediary, even in the case where, under the Convention law, the rights resulting from the credit of those securities to a securities account are determined to be contractual in nature (see Examples 2-4 to 2-6 and comments in para. 2-17).

2-31 The English and French texts of Article 2(2) do not match. While the English text correctly refers to both a "disposition of" and "an interest in" securities, the French text only refers to a disposition (of either securities or an interest in securities). The history and purpose of Article 2(2) clearly reveal that the English version is correct and that the French text must be read accordingly.

IV. Article 2(3)(a) and (b): Subject to Article 2(2), the Convention does not determine the applicable law to purely contractual or otherwise purely personal rights between the parties to an account agreement or a disposition

2-32 With respect to the issues specified in Article 2(1), the Convention determines the law that governs the legal nature and effects against the intermediary and third parties of the rights resulting from a credit of securities to a securities account and of a disposition of securities held with an intermediary. Article 2(3)(a) and (b), which is subject to Article 2(2), reiterates that, with respect to issues not covered by Article 2(1), the Convention does not determine the law applicable to purely contractual or otherwise purely personal rights and duties of the parties to an account agreement or a disposition *inter se*. For example, the Convention does not determine the law applicable to matters (often but not necessarily dealt with in the account agreement) such as the intermediary's standard of care in maintaining the securities account and any securities credited to it, the content and frequency of account statements, the deadlines for when the account holder must give instructions to ensure they are carried out on the same day, or the risk of loss (*e.g.*, computer failure) of securities held for the account holder, as between the account holder and the intermediary (Art. 2(3)(a)). Similarly, the Convention does not determine the law applicable to matters relating to a disposition such as the number and type of securities to be disposed, the price for the securities, the date on which the securities must be transferred against payment, or the consequences of a failure by either party to transfer securities or make payment when contractually obligated to do so (Art. 2(3)(b)).

2-33 The law applicable to those contractual rights between an account holder and its intermediary *inter se* which fall outside Article 2(1) is determined not by the Convention but by the other conflict of laws rules of the forum. These rules typically allow the parties to an account agreement or to a collateral or other disposition agreement to choose the law governing their contractual rights and duties under that agreement. Article 4(1), discussed below, in providing the conflict of laws rule governing the Article 2(1) issues, explicitly gives effect to the choice made by the parties to an account agreement to govern the Article 2(1) issues without requiring the same choice to be made (or indeed that any choice at all be made) with respect to the law governing other issues, including the types of purely contractual or purely personal rights described above. Likewise, nothing in the Convention precludes the parties to a disposition agreement from choosing the law to govern other matters relating to that agreement. Such a choice, however, will have no effect on the applicability of the Convention law with respect to the Article 2(1) issues.

Example 2-11

A customer, C, has securities credited to a securities account maintained for it by an intermediary. C transfers the securities to a third-party, T, a customer of the same intermediary. C and the intermediary expressly agree on (i) the intermediary's liability for acts or omissions in maintaining securities accounts and any securities credited to them being limited to liability for wilful misconduct or gross negligence, (ii) the content and timing of the account statements that the intermediary will send to the customer, (iii) what the deadline is for the customer to give instructions to ensure that they are carried out on the same day and (iv) who bears the risk of loss of securities held for the customer under a variety of circumstances. Similarly, C and T expressly agree on (i) the number and type of securities to be disposed, (ii) the price to be paid for the securities, (iii) the date on which the securities must be transferred against payment and (iv) the consequences of a failure by either party to transfer securities or make payment when contractually obligated to do so. Later, in a proceeding in Ruritania, where the Convention is in force, the issue is whether the Convention determines the law applicable to determine the interpretation or effectiveness of any of these agreements. The Convention does not determine the law applicable to any of these matters because none of them falls within any of the issues specified in Article 2(1) relating to the securities. The above matters do, however, fall within Article 2(3)(a) and (b) and the law applicable to them is determined by the conflict of laws rule of the forum other than those contained in the Convention.

V. Article 2(3)(c)

2-34 The Convention does not determine the law applicable to the rights and duties of an issuer of securities or of an issuer's registrar or transfer agent, whether securities are directly held or held with an intermediary (see Examples 2-1 and 2-2). This exclusion encompasses the duties of an issuer with respect to all corporate actions, including voting rights, dividend rights and registration rights, and the rights of an issuer to define the steps for achieving good discharge of a note, bond or other debt security.

Example 2-12

A company organised under the laws of England issues securities represented by a global certificate, which is registered in the name of the nominee of a central securities depository (CSD). Account holders of the CSD are permitted to enforce the issuer's payment obligations under the securities directly against the issuer in the event of the issuer's default on the securities under the terms of the securities or a separate deed of covenant. The issuer defaults on the securities. Even though the account holders have the right to enforce the securities directly against the issuer in the event of default, the securities are still treated as securities held with an intermediary. Therefore, the Convention determines the law governing the Article 2(1) issues related to an interest in or a disposition of the securities. In any event, the Convention does not determine the law governing the issuer's rights and duties under the securities.

2-35 It should be noted also that the Convention has no impact on regulatory rules relating to the issue or trading of securities.

Article 3 *Internationality*

This Convention applies in all cases involving a choice between the laws of different States.

I. Introduction

3-1 Article 3 ensures the applicability of the Convention whenever a situation involving securities held with an intermediary relates in any way to more than one State ('internationality'). As any foreign element triggers the applicability of the Convention, any national conflict of laws rule on Article 2(1) issues which might otherwise be applicable is displaced by the Convention's universally applicable conflict of laws rules; this provides legal certainty and predictability as to the applicable conflict of laws rules and thereby as to the applicable substantive legal system.

3-2 The purpose of broad applicability of the Convention reflected in Article 3 has been uncontested from the very outset of the proceedings.

3-3 Article 3 does not define the applicability of the Convention by reference to specific, pre-established and precisely delimited factors or by means of a definition of "internationality" which would have to be satisfied at a particular point in time and against which parties and courts would have to assess facts to determine whether or not the Convention applies. Rather, the provision takes a broad descriptive approach by simply stating that the Convention applies in *all* cases involving "a choice between the laws of different States". Thus, the Convention applies unless there is absolutely no element in the facts of a case (*e.g.*, 'location' of a person involved in or affected by a transaction or of an activity of such a person, 'location' of a security or its issuer, presence of a governing law clause or any other 'governing law' factor or element) that might require a decision as to which of two or more legal systems is applicable. If there were indeed absolutely no element that might require such a decision, the mere fact that the forum is a "foreign" State would not by itself require such a decision. Most importantly, the factors mentioned in Article 6 which are to be disregarded *in determining the applicable law*, remain relevant to the question whether the situation under consideration involves a foreign element for the purpose of Article 3. Only a formulation that does not depend on the development of a coherent interpretation of Article 3 in all States Party, but instead declares a broad scope of application that cannot be avoided by (mis)interpretation of the words used, enables parties to a disposition and affected third persons to enjoy the legal certainty and predictability as to the governing law that the Convention seeks to provide.

3-4 The applicability of the Convention does not require the existence of a "conflict of laws" under the private international law rules of the forum or any other State. Thus, the fact that the forum might not consider the foreign element to be material to the particular issue before it, is irrelevant for the applicability of the Convention. For example, where there is a priority dispute between X and Y relating to securities, issued by Z being incorporated under the laws of England, held by I in an account with B and successively given in security to X and Y, and B, I, X and Y are all in Spain, the fact that the issuer has nothing to do with the priority dispute does not remove the foreign element which triggers the applicability of the

Convention. The text of Article 3 does not use the term 'internationality' as a presupposition of the Convention's applicability; the term is used only in the title of the Article to alert a reader to the general content of the Article. The choice not to use 'internationality' in the Convention text was deliberate; invocation of that term might cause the exclusion of many cases that are intended to be covered by the Convention. As discussed below, many situations that appear, at first glance, to be wholly 'internal' are covered by the Convention.

3-5 The reference to a "choice between the laws of different States" does not mean that the Convention applies only to situations where contracting parties have 'chosen' - in the sense of 'agreed upon' - the law governing their agreement (be it the account agreement, the agreement relating to a disposition or any other agreement). Nor does it mean that a traditional conflict of laws analysis under the forum's private international law must be engaged in or satisfied. The expression 'choice' refers to a determination of applicable law that is made with respect to an Article 2(1) issue because a situation involving securities held with an intermediary includes elements that relate in any way to different States, thereby giving rise to the question as to which of the potentially applicable laws should govern the issue. Any situation implying the possible applicability of the law of different States triggers application of the Convention, with the consequence that the applicable law will be determined by the Convention's conflict of laws rules.

II. Illustrations of the applicability of the Convention

3-6 The Convention applies to all situations involving securities held with an intermediary which relate in any way to more than one State. The 'foreign' element of a situation may derive from one of the parties, from a governing law clause, or from any other element.

3-7 It might arise, for example, with respect to any of the persons identified in the following list which is however *by no means exhaustive*:

- the account holder;
- any of the parties to a disposition of the securities or the securities account, or an interest in either;
- the relevant intermediary; or
- an issuer (see para. 3-8 and Example 3-2).

Whenever any of these persons has, for example, its place of business or incorporation, or its habitual residence or domicile, in a different State, or any of these persons acts in a different State in a situation involving securities held with an intermediary, the Convention applies.

Example 3-1

> An account holder who is an individual habitually resident in Brazil opens a securities account with an intermediary organised under the laws of Brazil. All the securities credited to the securities account are securities issued by issuers organised under Brazilian law or by the Brazilian government. The account agreement expressly provides that it is governed by the laws of Brazil and does not expressly provide that a different law governs all the issues specified in

Article 2(1). At the time the agreement on governing law was entered into, the intermediary had a Qualifying Office (Art. 4(1), second sentence) in Sao Paolo. Later, the account holder grants a security interest in the securities account and all securities credited to the securities account to a Bank organised under the laws of Mexico. In a proceeding in a State where the Convention is in force (whether Brazil, Mexico or elsewhere), the issue is whether the security interest is perfected. The Convention applies because the situation concerns securities held with an intermediary and involves elements that relate to more than one State (*i.e.*, to Brazil and Mexico). This, in the language of Article 3, produces a situation necessitating a "choice between the laws of different States". Hence, under Article 3, the Convention applies.

Example 3-1 illustrates not only the fact that the 'location' of the secured party, the transferee of the disposition, can provide the 'foreign' element, but also the broader point that although the opening of the securities account and the crediting of the securities are wholly domestic to Brazil and at that moment in time no 'foreign' element appears to be present, this fact does not preclude the Convention becoming applicable by reason of the disposition to the Mexican Bank.

3-8 The applicability of the Convention may also be triggered by an element relating to any issuer of securities credited to the securities account, any intermediary through whom the relevant intermediary holds the securities, or any other intermediary involved in a disposition of the securities (see Example 3-2 below). This reference to issuers and higher-tier intermediaries must not be misunderstood to suggest a "look-through" approach in determining a conflict of laws rule or in applying the Convention (on the firm and unambiguous rejection of the 'look-through approach' for conflict of laws purposes and the reasons for it, see comments in paras. Int-33 *et seq.*, in particular paras. Int-38 *et seq.*); the reference only indicates that even these persons, who play no role under the Convention in determining the applicable law, may nevertheless trigger the applicability of the Convention, (*i.e.*, they may constitute the element that relates the situation to more than one State).

Example 3-2

An account holder organised under the laws of Norway opens a securities account with an intermediary that is also organised under the laws of Norway. The securities credited to the securities account are issued by issuers organised under Norwegian, Finnish and Swedish law. The account agreement expressly provides that it is governed by the laws of Norway and does not expressly provide that a different law governs all the issues specified in Article 2(1). At the time the agreement on governing law was entered into, the intermediary was engaged in a business of maintaining securities accounts through an office in Oslo. Later, the account holder grants a security interest in the securities account and all securities credited to the account to a Bank organised under the laws of Norway. In a proceeding in a State where the Convention is in force, the issue is whether the security interest is perfected. The Convention applies because the situation involves securities held with an intermediary and relates to more than one State (*i.e.*, to Norway, Finland and Sweden). This creates a situation which involves a "choice between the laws of different States". Hence, under Article 3, the Convention applies.

Example 3-2 illustrates that even when it is only a relatively minor element (*e.g.*, only a few non-Norwegian-issued securities) that relates the situation to another State, and even when the situation is otherwise largely one-State-focused, that element is sufficient to trigger the applicability of the Convention. It is obvious that making the applicability of the Convention depend on the magnitude of the non-Norwegian-issued securities would not only be arbitrary, but would also necessitate a very detailed subsequent factual inquiry and detract from the certainty that the Convention seeks to provide.

3-9 The element relating the situation to another State is not limited to the 'location' of a person or its activities; other elements, such as the agreement between an account holder and its intermediary on governing law designating the law of that other State or (if it is a Multi-unit State) of a territorial unit of that other State, may trigger the applicability of the Convention (see the following Example 3-3), independently of whether or not the Qualifying Office requirement (see Art. 4(1), second sentence) is fulfilled.

Example 3-3

An account holder incorporated under the laws of Romania opens a securities account with an intermediary that is also incorporated under the laws of Romania. All securities credited to the securities account are issued by issuers incorporated under the laws of Romania. The account agreement expressly provides that it is governed by the laws of England and does not expressly provide that a different law governs all the issues specified in Article 2(1). At the time the agreement on governing law was entered into, the intermediary had a Qualifying Office (Art. 4(1), second sentence) in London (England). Later, the account holder grants a security interest in the securities account and all securities credited to the account to a Bank organised under the laws of Romania. In a proceeding in a State where the Convention is in force, the issue is whether the security interest is perfected. The Convention applies because the situation involves securities held with an intermediary and relates to more than one State (*i.e.*, to Romania and England). This produces a situation involving a "choice between the laws of different States". Hence, under Article 3, the Convention applies.

Example 3-3 makes clear that in cases where "all the factors relevant to the situation are connected with a single State, but the law agreed in the account agreement under Article 4(1) is the law of another State or of a territorial unit of (another) Multi-unit State" (see wording of Art. 3(b) Working Doc. No 19 of the Diplomatic Conference), the applicability of the Convention is triggered.

3-10 The Diplomatic Conference expressly discussed whether in such a case (Example 3-3 above) the *principle* of 'fraud of law' ('*fraude à la loi*', '*fraus legis*') or similar principles (*e.g.*, under English law: '*bona fide*') could be relevant with the consequence of displacing the applicability of the Convention triggered by the parties' choice. There was a clear consensus that such agreements cannot be treated as fraud cases and therefore cannot be rejected on the basis of the "fraud" argument. Otherwise a court of a State where the

Convention is in force could, by using the principle of 'fraud of law', prevent application of the Convention and apply its own national conflict of laws rules despite the fact that the Convention has been accepted by its own State as a universal conflict of laws regime replacing its national conflict of laws rules. Additionally, such a decision would encroach upon the declared common will of the parties concerned. Exercise of such party autonomy is contemplated and endorsed by the Convention. Furthermore, any possible fraud by the parties in agreeing on the governing law (with the consequence under the Convention of the designation of the applicable law) is prevented by the Qualifying Office requirement in Article 4(1) (see comments in para. 4-7). In addition, fraud between the account holder and its intermediary should not in any event affect innocent third parties. Applicability of the fraud of law principle would undermine the ability of third parties to rely on the Convention.

III. Applicability – Time

3-11 The applicability of the Convention (as distinguished from its entry into force) may commence at any point in time. Article 3 ensures that the Convention's application is not limited temporally to the moment of litigation or to the moment of the occurrence of a particular disposition or crediting to a securities account. Rather, the occurrence of circumstances at a point in time after a particular transaction (*e.g.,* the subsequent acquisition of rights by an adverse claimant) will also make the Convention applicable. Thus, prudent transacting parties will always consider the possibility that the applicability of the Convention may already have been triggered or may thereafter be triggered, in addition to the possibility that it will be applicable by reason of the circumstances of their particular transaction. This likely prudent behaviour also supports the approach that Article 3 should be given the widest possible interpretation (since applicability of the Convention is very unlikely to come as a surprise to anyone).

3-12 The reference in the English text of Article 3 to "cases" does not mean that the Convention applies only in a judicial proceeding. In fact, the reference has to be understood in the sense of 'situations' (see French text of Art. 3). The primary purpose of the Convention is to provide *ex ante* certainty to transacting parties, and thus the Convention applies (in the sense of 'has to be taken into account') whenever an analysis of governing law is being made, outside of (as well as inside) a judicial proceeding.

IV. Conflicts between territorial units within Multi-unit States

3-13 Most of the other modern Hague Conventions[23] and many other international treaties[24] contain a provision stating that a Multi-unit State shall not be bound to apply the rules of the Convention to conflicts having to do solely with two or more territorial units of

---

[23] See, *e.g.*, Art. 20 of the *Convention of 14 March 1978 on the Law Applicable to Agency*; Art. 20 of the *Convention of 22 December 1986 on the Law Applicable to Contracts for the International Sale of Goods*; Art. 21 of the *Convention of 1 August 1989 on the Law Applicable to Succession to the Estates of Deceased Persons*; Art. 44 of the *Convention of 13 January 2000 on the International Protection of Adults*.

[24] See, *e.g.*, Art. 19(2) of the *Rome Convention of 1980 on the Law Applicable to Contractual Obligations* ('Rome I'), and Art. 24 of the *Inter-American Convention of 1994 on the Law Applicable to International Contracts*.

that State or with that State itself and one or more of its territorial units. It was *not* felt necessary to include such a provision in the Securities Convention. First, Article 3 states that the Convention applies in situations concerning the laws of "different States", thereby making clear that the Convention does not apply to purely internal conflicts. More importantly, however, in the light of the characteristics of modern indirect holding patterns and the broad applicability of the Convention ensured by Article 3 (including the fact that the relation to another State may arise after the occurrence of a particular disposition), there is very little room for a purely internal situation to arise. There is nothing that prevents a Multi-unit State from applying, as a matter of purely internal law, the rules set forth in the Convention to conflicts that have to do solely with two or more territorial units of that State or with that State itself and one or more of its territorial units.

3-14 Article 12 ("Determination of the applicable law for Multi-unit States") does not enlarge the scope of the Convention to conflict of laws situations within a Multi-unit State (*i.e.*, "a State within which two or more territorial units of that State, or both the State and one or more of its territorial units, have their own rules of law in respect of any of the issues specified in Article 2(1)"; see comments on Art. 1(1)(m) in para. 1-28). Rather, Article 12 presupposes the applicability of the Convention under Article 3, that is, a conflict situation between the laws of different States, and provides interpretative and substantive provisions relating to the application of the Convention with regard to Multi-unit States (see comments on Art. 12).

CHAPTER II APPLICABLE LAW

Article 4 *Primary rule*

1. The law applicable to all the issues specified in Article 2(1) is the law in force in the State expressly agreed in the account agreement as the State whose law governs the account agreement or, if the account agreement expressly provides that another law is applicable to all such issues, that other law. The law designated in accordance with this provision applies only if the relevant intermediary has, at the time of the agreement, an office in that State, which –

a) alone or together with other offices of the relevant intermediary or with other persons acting for the relevant intermediary in that or another State –

i) effects or monitors entries to securities accounts;

ii) administers payments or corporate actions relating to securities held with the intermediary; or

iii) is otherwise engaged in a business or other regular activity of maintaining securities accounts; or

b) is identified by an account number, bank code, or other specific means of identification as maintaining securities accounts in that State.

2. For the purposes of paragraph (1)(a), an office is not engaged in a business or other regular activity of maintaining securities accounts –

a) merely because it is a place where the technology supporting the bookkeeping or data processing for securities accounts is located;

b) merely because it is a place where call centres for communication with account holders are located or operated;

c) merely because it is a place where the mailing relating to securities accounts is organised or files or archives are located; or

d) if it engages solely in representational functions or administrative functions, other than those related to the opening or maintenance of securities accounts, and does not have authority to make any binding decision to enter into any account agreement.

3. In relation to a disposition by an account holder of securities held with a particular intermediary in favour of that intermediary, whether or not that intermediary maintains a securities account on its own records for which it is the account holder, for the purposes of this Convention –

a) that intermediary is the relevant intermediary;

b) the account agreement between the account holder and that intermediary is the relevant account agreement;

c) **the securities account for the purposes of Article 5(2) and (3) is the securities account to which the securities are credited immediately before the disposition.**

I. Introduction

4-1 Article 4 sets forth the primary conflict of laws rule to determine the law applicable to the issues specified in Article 2(1) (the "Article 2(1) issues"), namely the law in force in the State (or territorial unit of a Multi-Unit State) expressly agreed in the account agreement as the State (or territorial unit of a Multi-Unit State) whose law governs the account agreement or, if the account agreement expressly provides for the law of a particular State (or territorial unit of a Multi-Unit State) to govern the Article 2(1) issues, then that law. If the law of a particular State is expressly agreed upon in an account agreement to govern all the Article 2(1) issues, that law will govern those issues whether or not the account agreement specifies the law by which the agreement itself is to be governed. But the Convention's primary rule applies only if, at the time of the agreement (by which is meant the agreement on governing law), the relevant intermediary has in the selected State an office (a "Qualifying Office") which (a) alone or with another office or third party wherever situated is engaged in a business or other regular activity of maintaining securities accounts (though not necessarily the particular account in question), whether in the form of an activity specified in Article 4(a)(i) or (ii) or otherwise, or (b) is identified by an account number, bank code or other specific means of identification as maintaining securities accounts in the selected State. Article 4(2) contains a list of functions none of which, standing alone, would amount to the functions required under the condition specified in Article 4(1)(a)(iii). If the applicable law is not determined under Article 4, one of the fall-back rules in Article 5 will determine the applicable law. Article 6 sets forth a list of factors that are not permitted to be considered in determining the applicable law under the Convention.

4-2 Article 4 determines the Convention law regardless of whether the account holder's interest in the securities is classified as proprietary, contractual, mixed or other, or whether the account holder has a right to enforce the securities directly against the issuer.

4-3 To allow the parties to an agreement to determine the law governing the rights of third parties and priority between competing claimants is in many jurisdictions a departure from traditional conflict of laws rules by which property rights and dispositions are governed by the *lex rei sitae*. However, the extension to intangible property of a concept devised for tangibles has proved of doubtful utility, involving the attribution of an artificial location to an asset which by its nature has no physical manifestation. Moreover, given that the location ascribed to an intangible varies according to its nature – in particular, whether it is a debt, a registered security or a bearer security – the *lex rei sitae* (more specifically referred to in the context of securities as the *lex cartae sitae*) does not constitute an organising concept and is best discarded in favour of a direct rule. Attempts to ascribe an artificial *situs* to securities held with an intermediary have proved unsatisfactory. For directly held registered securities a rule commonly applied is the place of the issuer's incorporation, or alternatively the place where the issuer maintains its securities register, and for directly held bearer securities the location of the certificates representing them. But these rules have resulted in legal uncertainty, unnecessary costs and serious practical difficulties when applied to securities held by an investor in an account maintained by an intermediary, particularly since a single securities account may contain securities issued by a variety of issuers organised under the laws of, or represented by certificates located in, a variety of States, and since there may be several tiers of intermediaries between each issuer and the ultimate investor. Even those who initially supported an attempt to extend the *lex rei sitae* rule to securities held with an intermediary were unable to develop a test for determining the location of a particular securities account that worked reliably in all or even most States.

4-4 These difficulties with the *lex rei sitae* approach led to the proposal to adopt the Place of the Relevant Intermediary Approach (PRIMA), that is, the law of the place where the account holder's direct – or relevant – intermediary maintains the securities account for the account holder, which at first seemed to provide a much more realistic connecting factor. However, it became apparent in the course of the negotiations that the Convention had to move beyond the initial formulation of the PRIMA principle because of the difficulty of identifying the relevant location of the intermediary, given that in modern global trading the various activities involved in the maintenance of securities accounts are often dispersed among offices in different countries, shift frequently, are often outsourced, and are becoming increasingly electronic so that, for example, while the investor may sign the account agreement in country A the account may be controlled from country B and the intermediary's electronic records centralised in country C (see paras. Int-41 *et seq.*, and 4-24 to 4-26).

4-5 In the light of these problems with the *lex rei sitae* and the initial formulation of the PRIMA approaches, it was decided to move beyond the initial formulation of the PRIMA principle. This was done in two ways: (i) the focus on attributing a "location" to an intermediary or a securities account was abandoned and replaced with an approach giving effect to an agreement on governing law between an account holder and its intermediary and (ii) a Qualifying Office requirement was added. This approach has several advantages. First, it provides certainty, in contrast to the present position where the applicable law as determined by the conflict rules of the forum may vary from jurisdiction to jurisdiction, leading to unpredictability and the prospect of forum shopping to select the forum whose conflict rules will lead to the application of the law most favourable to the claimant's case. Secondly, it routes all the Article 2(1) issues with respect to a particular securities account to the same law, thereby obviating the risk that different parts of a single transaction or a connected series of transactions with respect to that securities account will be governed by different laws. A party intending to acquire a security interest in securities credited to a securities account can always ask for a copy of the account agreement and information as to the state of the account if it feels the need for this. In moving beyond the initial formulation of the PRIMA principle, the Convention retains for the purposes of its conflict rules the concept of the relevant intermediary and rejects any *lex rei sitae* or "look-through" approach or any "Super-PRIMA" (see paras. 4-11 and 4-43 *et seq.*).

4-6 The primary rule is based on a relationship between an account holder and its intermediary and avoids the need for any attempt (real or artificial) to locate a securities account, the office that maintains a securities account, the underlying securities or the records of the issuers of the underlying securities. It also does not look to the place where the parties to the account agreement or other persons reside, are organised, conduct business generally, entered into the account agreement or otherwise took action related to the securities account or the securities. Equally irrelevant is the fact that under the substantive rules of the *lex fori* or any other law an account holder is entitled to look through his own intermediary and assert rights directly against a higher-tier intermediary or the issuer. The primary rule leaves it to the parties to the account agreement[25] to select the law governing all the Article 2(1) issues with respect to the securities credited to the account. It reflects a policy decision to formulate a conflict of laws rule that reduces risk, promotes capital formation and investment, reflects existing and foreseeable market practice, is practical and efficient, and permits market participants to determine in advance with respect to securities held with an intermediary which law governs all the Article 2(1) issues, providing *ex ante* legal certainty and predictability for the largest number of situations and transactions.

4-7 Nevertheless, the Diplomatic Conference recognised the concerns expressed by various delegations about allowing the parties to an account agreement complete freedom to choose the law governing the Article 2(1) issues. These concerns were addressed by a provision rendering ineffective the parties' choice of the law of a State in which the intermediary has no Qualifying Office at the time of the agreement on governing law. See comments in paragraphs 4-21 *et seq.*

4-8 To prevent courts from applying traditional concepts such as the law of the place of incorporation of the issuer or the law of the location of certificates to securities held with an intermediary, Article 6 lists these and other factors which are to be disregarded in determining the Convention law.

4-9 If the applicable law is not determined under Article 4, either because the parties have expressly selected an applicable law but the Qualifying Office requirement is not satisfied or because they have failed to make any express selection in their agreement or have never concluded an account agreement at all (see para. 5-7), the fall-back rules of Article 5 will be applied to determine the law governing all the Article 2(1) issues. These are arranged in a cascade. Under the first fall-back rule, if a written account agreement contains an express and unambiguous statement that the relevant intermediary entered into the account agreement through a particular office, the applicable law is the law in force in the State in which that office was located at the time the account agreement was entered into, provided that the office is a Qualifying Office (on the Qualifying Office requirement, see paras. 4-21 *et seq.*; on the first fall-back rule, see paras. 5-4 *et seq.*). Since the relevant office for the purpose of the first fall-back rule is not the same as the relevant office for the purpose of Article 4(1), it is quite possible to have a situation in which the Qualifying Office requirement is not satisfied under Article 4(1) but is satisfied under Article 5(1). If the account agreement is not

---

25 It is the agreement between those two parties that is given regard under the Convention, even in the context of a disposition – where the account holder disposes of securities that have been credited to his securities account the effects of the disposition are governed by the applicable Convention law chosen by the parties to the account agreement (or, in default, by one of the default rules in Art. 5), and it is not open to the parties to the disposition to subject its effects to a different law by a choice of law clause. The Convention law applies not only to the relations between the account holder and its intermediary but also to the effects of any disposition of the securities by the account holder (see commentary on Art. 2).

in writing or it contains no such statement or if the office is not a Qualifying Office the second fall-back rule is applied, namely the law in force in the place under whose law the intermediary was incorporated or otherwise organised at the time of entry into the account agreement or, if there is no written agreement, at the time of opening of the securities account (see paras. 5-7 *et seq.*). Where this intermediate fall-back rule does not apply, because the intermediary was not at the time of the agreement a corporation or an otherwise organised entity, the ultimate fall-back rule comes into play, namely the intermediary's place of business, or if it has more than one then its principal place of business, at the time the written account agreement was entered into, or, if there is no written agreement, the securities account was opened (see para. 5-11).

4-10 It is important to note that the applicable law determined under the Convention applies to *all* of the Article 2(1) issues. Hence, it is not possible that some of these issues will be governed by one law while others are governed by a different law, nor is it open to the parties to select the law to govern only some of the Article 2(1) issues while leaving the remaining issues to be dealt with under Article 5. Such a provision would be ineffective and would lead to the application of the law expressed to govern the account agreement or, if there was no or no effective choice of such law, then the relevant fall-back rule in Article 5.

4-11 The Convention rules apply separately with respect to each securities account (in other words, to each relationship between an account holder and its relevant intermediary). Thus, when a chain of intermediaries stands between an account holder and the issuer or between an account holder making a disposition and the counterparty of that disposition, there is no single law, such as the place of the issuer's incorporation, which would govern all the Article 2(1) issues with respect to all securities accounts maintained by intermediaries standing between the account holder and the issuer or between the account holder and its counterparty. Any such result would be incompatible with the fundamental principle underlying the Convention that the law applicable should be that which is based on the relationship between a particular account holder and its relevant intermediary (*i.e.*, the particular securities account and the rights resulting from a credit of securities to it) (see also paras. Int-30 and Int-31). Application of Articles 4 and 5 in the context of a disposition of securities effectuated by transfer from one securities account to another, including through a chain of intermediaries, is discussed at length and illustrated by useful examples in paragraphs 4-43 *et seq.*

4-12 Article 4(3) sets forth interpretive rules with respect to a disposition by an account holder of securities held with an intermediary in favour of that intermediary (see paras. 4-41 and 4-42).

4-13 Articles 4 and 5 must be read in conjunction with Article 7 (protection of rights on change of law resulting from an amendment to an account agreement), Article 12 (determination of the applicable law for Multi-unit States) and Article 16 (Pre-Convention account agreements and securities accounts). See the commentaries to these Articles.

II. The primary rule: Article 4(1), first sentence

4-14 Subject to the Qualifying Office requirement in the second sentence (see paras. 4-21 *et seq.*), the primary rule, formulated in the first sentence of Article 4(1), is as follows:

(1) If an account holder and its relevant intermediary expressly agree that the law of a particular State will govern their account agreement, that law will govern the Article 2(1) issues (Example 4-1), unless the parties have in addition expressly selected a different law to govern all the Article 2(1) issues, in which case that law would apply;

(2) if the parties do not expressly agree on a law to govern the account agreement but do agree on the law of a particular State to govern *all* the Article 2(1) issues, then that law will govern those issues.

4-15 The phrase "law in force" in Article 4(1) (and Art. 5) is used in preference to "law of" to cover cases where, in a territorial unit, the relevant law includes both the law of that territorial unit and, to the extent applicable (under the law of either the unit or the Multi-Unit-State) in that unit, the law of the Multi-unit State.[26]

4-16 If the parties agree on a law to govern only one or some of the Article 2(1) issues, that choice is not effective under Article 4(1) to determine the applicable law. In such a case, then if the parties have also expressly agreed on a law to govern the account agreement generally and assuming the Qualifying Office requirement is satisfied, the law expressly selected by them to govern the account agreement will, under Article 4(1), determine the applicable law (Example 4-2); if the Qualifying Office requirement is not satisfied, the relevant fall-back rule in Article 5 will determine the applicable law.

4-17 Under Article 4(1), the law applicable to the Article 2(1) issues must be determined on the basis of what the parties have *expressly* agreed in the account agreement. The choice of a law to govern the Article 2(1) issues may under no circumstances be implied from the terms of the account agreement considered as a whole or from the surrounding circumstances (see Example 4-7).[27] This is because it was felt that the possibility for a court to imply an unexpressed choice would reduce the certainty of the conflict of laws regime created by Articles 4 and 5. An implied choice of law under the same account agreement may still be effective to govern matters that are outside the scope of Article 2(1) and hence the Convention (see commentary to Art. 2, in particular paras. 2-4 to 2-8), but the Article 2(1) issues will be governed by the law applicable under the relevant fall-back rule in Article 5 and not by any other law the parties may impliedly have chosen.

---

26 On the operation of the primary rule in relation to Multi-unit States, see the commentary to Art. 12; Art. 12(2)(a) expressly clarifies the meaning of the expression "law in force in a territorial unit of a Multi-unit State", which is used in several provisions of the Convention (see paras. 12-12 and 12-13).

27 This is in contrast to Art. 16(4), which, in the limited circumstances in which that provision is applicable, gives a Convention effect to an agreement between the parties as to where the securities account is maintained even when that agreement is implied from the terms of the account agreement or from the surrounding circumstances at the time the account agreement was entered into (see paras. 16-20 and 16-21). The rules of Art. 16 for interpretation of pre-Convention account agreements which do not expressly refer to the Convention have no bearing on the meaning or effect of language in post-Convention agreements or those pre-Convention agreements which expressly refer to the Convention (see Art. 16(2)), and there is no logic or policy that would compel or even support conclusions based on *a contrario* arguments.

4-18 The choice of law must be expressed as part of the account agreement. The account agreement may consist of more than one document. An express choice of law provision in the general terms and conditions of the account agreement is sufficient under Article 4(1). There is no requirement in Article 4(1) that the account agreement be in writing; indeed, writing is not required anywhere by the Convention, except under Article 5. So an oral express choice of law is effective.[28] It is only the choice of law itself that need be express. The fact that there are other terms which are merely implied is irrelevant. The time at which the Qualifying Office test is to be satisfied and the effect of a modification of the account agreement which changes or singles out and expressly reaffirms an agreement on the governing law are discussed in paragraph 4-27.

Example 4-1

An intermediary and its customer enter into an account agreement which expressly provides that it is governed by the laws of Singapore and does not expressly provide that a different law governs all the Article 2(1) issues. At the time the agreement on governing law was made, the intermediary was engaged in a business of maintaining securities accounts through an office in Singapore as described in Article 4(1)(a)(iii). Later, the customer grants a security interest in the securities account and all securities credited to the account. The secured party perfects its interest in accordance with Singapore law. Still later, in a proceeding in a State where the Convention is in force, the issue is whether the security interest is perfected. Under Article 4(1), Singapore law governs the Article 2(1) issues, including the perfection issue, because Singapore law was expressly agreed as the law governing the account agreement, another law was not agreed to govern all the issues specified in Article 2(1), and the Singapore office was a Qualifying Office at the time the agreement on governing law was made (see paras. 4-21 *et seq.*). Even if it were possible, it is not necessary to determine whether the securities account was at any time maintained by the intermediary at the Singapore office; that fact would be irrelevant.

Example 4-2

The facts are as in Example 4-1 except that the account agreement provides that Italian law governs the perfection issue but not all the Article 2(1) issues and the secured party perfects its interest under Italian law. Article 4(1) will not give effect to the agreement on Italian law because it was not agreed upon for all the Article 2(1) issues. Under Article 4(1), Singapore law governs all the Article 2(1) issues, including the perfection issue, because Singapore law was expressly agreed as the law governing the account agreement, the selection of Italian law to govern the perfection issue was not effective, and the Singapore office was a Qualifying Office at the time the agreement on governing law was made.

---

28 An oral governing law clause is effective for Convention purposes even if a writing or other formality requirement would render it ineffective under any private international law or substantive writing requirement.

4-19 The question whether an agreement on the governing law does not exist due to the absence of consent (*e.g.*, by reason of a generally applicable contract law doctrine such as lack of capacity) is governed by the conflict of laws rules of the forum other than those contained in the Convention. If, under the applicable substantive rule, consent is absent, there is in fact no agreement for the purpose of Article 4(1) and the relevant fall-back rule of Article 5 applies. If there is a consented-to agreement, a material rule depriving that agreement of legal effectiveness may be applied only in accordance with Article 11.

4-20 While all the Article 2(1) issues are required to be governed by the same law, it is open to the parties to the account agreement to select different laws for different securities held with the same intermediary by opening different accounts or sub-accounts, each governed by its own law, provided that in each case the Qualifying Office test is satisfied.

III. The Qualifying Office requirement: Article 4(1), second sentence, and Article 4(2)

A. Introduction

1. The conditions summarised

4-21 The effectiveness of the choice of governing law by the parties under the first sentence of Article 4(1), as well as the first fall-back rule in Article 5, is subject to the Qualifying Office requirement prescribed by the second sentence of Article 4(1). This requires that, at the time of the agreement on governing law (or, in the case of Art. 5(1), at the time the account agreement is first entered into), the relevant intermediary has an office which meets the requirements there set out. The effect of failure to satisfy the Qualifying Office requirement depends on the nature of the selection:

(1) If the parties choose the law to govern the account agreement but do not choose a different law to govern all the Article 2(1) issues, then if the choice fails to satisfy the Qualifying Office requirement it is necessary to resort to the applicable fall-back rule in Article 5.

(2) The same result follows in the converse case where the parties do not select a law to govern the account agreement but do select a law to govern all the Article 2(1) issues and the selection fails the Qualifying Office requirement.

(3) If the parties choose one law to govern the account agreement and a different law to govern all the Article 2(1) issues and the latter but not the former fails to satisfy the Qualifying Office requirement, the applicable law is that chosen by the parties to govern the account agreement. This is because the first sentence of Article 4(1) is predicated on the assumption that the choice of a different law to govern the Article 2(1) issues is an effective choice.[29]

---

[29] In the converse case where the Qualifying Office requirement is satisfied as to the law chosen to govern all the Article 2(1) issues but not as to the law chosen to govern the account agreement there is,

4-22 There are two separate conditions that must be met at the time of the agreement on governing law in order to satisfy the Article 4(1) Qualifying Office requirement:

(1) The relevant intermediary must have an office within the State whose law has been expressly agreed upon by the parties, an office as defined by Article 1(1)(j) (see paras. 1-22 *et seq.*).[30]

(2) That office must, either alone or with other offices of the relevant intermediary or other persons acting for it, be:

(a) engaged in a business or other regular activity of maintaining securities accounts, whether by fulfilling one of the functions specified in Article 4(1)(a)(i) or (ii) or otherwise; or

(b) identified by some specific means of identification as maintaining securities accounts in the State in question.

4-23 It is important to note that no aspect of the Article 4(1) Qualifying Office requirement relates to any *specific* account maintained by the relevant intermediary or any *specific* account holder. An office is a Qualifying Office if it engages in the business or other regular activity of maintaining securities accounts, or is identified by some specific means of identification as maintaining securities accounts in the selected State, even if the securities account of the particular account holder in respect of which an issue arises is maintained by an office situated in another State (assuming that one can determine where a particular account is maintained).

Example 4-3

An intermediary and its customer enter into an account agreement. The account agreement expressly provides that it is governed by the laws of Argentina and does not expressly provide that a different law governs all the Article 2(1) issues. The intermediary had an office in Buenos Aires at the time the agreement on governing law was made. A coding system identified the Buenos Aires office as maintaining securities accounts in Argentina as described in Article 4(1)(b). Later, the customer grants a security interest in the securities account and all securities credited to the account. The secured party perfects its security interest in accordance with Argentine law. Still later, in a proceeding in a State where the Convention is in force, the issue is whether the security interest is perfected. Under Article 4(1), Argentine law governs all the Article 2(1) issues, including the perfection issue, because Argentine law was expressly agreed as the law governing the account agreement, another law was not agreed to govern all the issues specified in Article 2(1), and the Buenos Aires office was a Qualifying Office at the time the agreement on governing law was made. Even if it were possible, it is

---

of course, no problem, since even if the latter choice had satisfied the Qualifying Office requirement effect would have been given to the choice of the former.

30 On the operation of the Qualifying Office requirement in relation to Multi-unit States, see the commentary to Art. 12.

not necessary to determine whether the securities account was at any time maintained by the intermediary at the Buenos Aires office or whether the Buenos Aires office ever engaged in maintaining securities accounts; those facts would be irrelevant.

Example 4-4

An intermediary organised under the laws of Panama and its customer enter into an account agreement. The account agreement expressly provides that it is governed by the laws of New York and does not expressly provide that another law governs all the issues specified in Article 2(1). The intermediary did not have an office anywhere in the United States at the time the agreement on governing law was made. Later, the customer grants a security interest in the securities account and all securities credited to the account. Still later, in a proceeding in a State where the Convention is in force, the issue is whether the security interest is perfected. The New York governing law clause does not have the effect under Article 4(1) of determining the applicable law because the intermediary did not have a Qualifying Office in the United States at the time the agreement on governing law was made. Consequently, the applicable law will be determined under Article 5. In this example, because Article 5 will be applied, the place of organisation of the intermediary is relevant.

Example 4-5

An intermediary and its customer enter into an account agreement. The account agreement expressly provides that it is governed by the laws of Portugal. It also expressly provides that Italian law governs all the Article 2(1) issues. At the time the agreement on governing law was made, the intermediary had an office in Lisbon which effected or monitored entries to securities accounts as described in Article 4(1)(a)(i); it did not have an office in Italy. Later, the customer grants a security interest in the securities account and all securities credited to the account. The secured party perfects its interest in accordance with Portuguese law. Still later, in a proceeding in a State where the Convention is in force, the issue is whether the security interest is perfected. Under Article 4(1), Portuguese law governs all the Article 2(1) issues, including the perfection issue. Because the intermediary did not have a Qualifying Office in Italy at the time the agreement that Italian law would govern all the Article 2(1) issues was made, that agreement does not have the effect under Article 4(1) of determining the applicable law. It also does not have the effect of displacing the choice of Portuguese law (*i.e.*, it does not preclude the general governing law clause from having its determinative effect under the primary rule). The choice of Portuguese law satisfies the condition because the intermediary had, at the time the agreement on governing law was made, a Qualifying Office in Portugal. The fact that the choice of Italian law does not determine the applicable law under the primary rule does not lead to the application of Article 5. This is because, under Article 4(1), a general governing law clause that satisfies the Qualifying Office condition is displaced only if there is also present an explicit choice of another law to govern all the issues specified in Article 2(1) and that explicit choice satisfied the Qualifying Office condition. Even if it were possible, it is not necessary to determine whether the securities account was at any time maintained by the intermediary at the Lisbon office; that fact would be irrelevant.

4-24 In today's market environment, it is increasingly the case that the activities involved in the maintenance and customer service operations relating to a single securities account may be undertaken from a variety of offices or outsourced to third parties in a multitude of locations. By way of illustration, assume an intermediary incorporated under the laws of the State of New York agreed with its client that the client's securities account is maintained in Tokyo because it is the place where the account was initially opened and where the first credit of securities to the account was effected. However, the intermediary sends all its account statements to the client from an office in Dublin. The client receives dividends administered and sent from an office in Hong Kong and obtains advice as to the ongoing status of the account from an office near the client's main office in Singapore. All the intermediary's operations relating to each of its securities accounts (including the entries effectuated) are backed-up and monitored by two separate computer systems run from its New Delhi and San Francisco offices respectively. Finally, the client regularly accesses the information relating to the relevant securities account from a laptop during many trips around the world.

4-25 In such a situation, if the test were the location of the securities account or the location of the office where the securities account is maintained, no certainty could be achieved and such a test would invite litigation in which courts would be required to make fact-intensive inquiries. The risks and burdens presented to a potential collateral taker are readily apparent.

4-26 The use of tax, regulatory reporting or accounting requirements to determine the location of a securities account or the office maintaining a securities account was also considered having regard that in certain States intermediaries may be required to assign a code to each securities account that effectively allocates it to a particular office for tax, regulatory reporting or accounting purposes, but this was rejected for a number of reasons. First, such tax, regulatory reporting or accounting requirements are by no means universal and may be the exception rather than the rule. Some States have comprehensive accounting rules for assets and liabilities that appear on an intermediary's balance sheet, but many do not have such rules for securities accounts that do not appear on the intermediary's balance sheet. Secondly, accounting, regulatory, and tax rules are based on considerations that are wholly unrelated to the considerations involving the global business of securities custody, clearing and settlement. Consequently, to use the allocation of a securities account to a particular office for tax, regulatory reporting or accounting purposes to determine the applicable law for an unrelated business purpose would be quite arbitrary. This is particularly true in light of the reality that some or all of the functions involved in the maintenance and servicing of a securities account are increasingly being undertaken from more than one office or outsourced to third parties in several locations.

Example 4-6

An intermediary and its customer enter into an account agreement. The account agreement expressly provides that Peruvian law governs all the Article 2(1) issues but that the account is to be maintained in Buenos Aires, where the intermediary had an office. The account agreement did not specify a law to govern the account agreement. At the time of the agreement the intermediary also had an office in Lima. There did not exist a specific means of identifying the Lima office as maintaining securities accounts. The Lima office did not, by itself, serve the functions described in Article 4(1)(a)(i) or (ii) or functions that amounted to the business of maintaining securities accounts as described in Article 4(1)(a)(iii). However, the functions served by the Lima office, when combined with the functions served elsewhere (not necessarily in Peru) by a third party to which the Peruvian intermediary had outsourced aspects of the maintenance of securities accounts, did amount to the business of maintaining securities accounts as described in Article 4(1)(a)(iii). Later, the customer grants a security interest in the securities account and all securities credited to the account. The secured party perfects its interest in accordance with Peruvian law. Still later, in a proceeding in a State where the Convention is in force, the issue is whether the security interest is perfected. Under Article 4(1), Peruvian law governs the Article 2(1) issues, including the perfection issue, because Peruvian law was expressly agreed as the law governing all the Article 2(1) issues, and the Lima office was a Qualifying Office at the time the agreement on governing law was made. Whether the securities account was at any time maintained by the intermediary at the Lima office is irrelevant. Even if it were possible, it is not necessary to determine whether the securities account was at any time maintained by the intermediary at the Lima office; that fact would be irrelevant. The parties' agreement as to the place of the maintenance of the account (*i.e.*, Buenos Aires) is equally irrelevant.

Example 4-7

An intermediary with an office in London enters into an account agreement in London with a United Kingdom company which has its head office in London. The agreement provides that it was entered into through the London office of the intermediary, that the account is to be maintained in London and that notices and other documents are to be served on the company at the address of its head office. The agreement does not contain any choice of law provision. The account is later credited with various securities acquired on the company's behalf, the issuers being incorporated in France, Germany and Italy. Subsequently the company pledges the French securities to a French bank under a pledge agreement concluded in Paris. A question arises whether the pledge was perfected. Although the circumstances might be sufficient to support a finding of an implied agreement by the intermediary and the United Kingdom company that the account agreement is to be governed by English law, Article 4(1) does not apply, there being no express agreement on the governing law, so that the question will be determined by the law applicable under Article 5(1), which in this example leads to the same result.

2. The time factor

4-27 The time when the condition in the second sentence of Article 4(1) has to be satisfied in order for there to be an office which is a Qualifying Office is the time when the express agreement on governing law is made. If at that time there is no office that is a Qualifying Office, the agreement on governing law is ineffective under Article 4(1) even if subsequently an office becomes a Qualifying Office. Conversely, if there is an office that is a Qualifying Office when the agreement on governing law is made the fact that it ceases to be subsequently or even is closed down does not affect the effectiveness of the governing law choice under Article 4(1). However, the Qualifying Office test must be reapplied whenever the account agreement is amended either by changing the governing law or by otherwise modifying the agreement and singling out and expressly reconfirming the preexisting governing law provision. A new application of the Qualifying Office test may lead to its being satisfied when previously it was not; it may on the other hand result in the Qualifying Office test ceasing to be satisfied because now the chosen law is that of a State where the intermediary does not have a Qualifying Office. In either case the effects of the modification operate only from the time it is made. There is no automatic re-application of (or retesting under) Article 4 each time an account agreement is amended (*e.g.*, a change in the schedule of fees); reapplication occurs only when the account agreement is amended either by changing the governing law or by otherwise modifying the agreement and singling out and expressly reconfirming the pre-existing governing law provision. See also the commentary on Article 7.

4-28 In virtually all cases, the agreement on governing law will be part of the account agreement *ab initio,* so the agreement on governing law will be made simultaneously with and be part of the account agreement. It is, though, possible that the parties might have entered into an account agreement without having included a governing law clause and, upon subsequently wishing to repair that omission, the parties might agree later on a governing law, that choice thereby becoming part of the account agreement. Or, having initially agreed on a governing law clause designating X law, the parties might later amend that provision of their account agreement to designate Y law as the governing law. In these instances, the Qualifying Office test has to be applied again in the light of the amendment (now applied as of the time of the amendment) and if it is then satisfied the choice of law or a different law is effective. This rule should be kept in mind also when dealing under Article 7 with respect to the effect of certain amendments to account agreements.

4-29 The timing rule under Article 4(1) should be contrasted with the timing rule for the Qualifying Office condition in the context of the fall-back rule in Article 5(1). The key element of that rule is the presence of an express and unambiguous statement in a written account agreement (relating to the intermediary's entering into that agreement "through a particular office"), rather than the existence of an agreement on governing law. Therefore, under Article 5(1), although the substance of the condition with respect to the existence of a Qualifying Office is the same as in Article 4(1), the pertinent point in time with respect to the satisfaction of that condition under Article 5(1) is "then" – which refers to the time the written account agreement was first entered into (not to the time of the making of an express agreement on governing law).

Example 4-8

An intermediary and its customer enter into an account agreement. The account agreement expressly provides that it is governed by the laws of England. It also expressly provides that Italian law governs all the Article 2(1) issues. At the time the agreement on governing law was made, the intermediary had an office in Milan at which it administered payments and corporate actions relating to securities as described in Article 4(1)(a)(ii). Later, the customer grants a security interest in the securities account and all securities credited to the account. The secured party perfects its interest in accordance with Italian law. Still later, the intermediary moves all of its functions relating to securities accounts to its office in Dublin. The Milan office thereafter performs no functions with respect to securities accounts. Much later, in a proceeding in a State where the Convention is in force, the issue is whether the security interest is perfected. Under Article 4(1), Italian law governs all the Article 2(1) issues, including the perfection issue. That is because the account agreement expressly specified Italian law to govern all the Article 2(1) issues, and, even though at the time of the proceeding the Milan office had ceased to be a Qualifying Office, it was a Qualifying Office at the time the agreement on governing law was made. The fact that the account was not maintained in that office is irrelevant. Even if it were possible, it is not necessary to determine whether the securities account was at any time maintained by the intermediary at the Milan office; that fact would be irrelevant.

3. "Office"

4-30 This is defined in Article 1(1)(j) (see paras. 1-22 *et seq.*).

4. The qualifying activity

4-31 Under Article 4(1)(a) the office must be one which:

(i) effects or monitors entries to securities accounts;

(ii) administers payments or corporate actions relating to securities held with the intermediary; or

(iii) is otherwise engaged in a business or other regular activity of maintaining securities accounts.

4-32 The last of these three heads of activity is not satisfied if the actual activity of an office consists merely of one of the activities listed in Article 4(2) (see para. 4-40). Article 4(2) sets forth a list of activities *none* of which, *standing alone*, is sufficient for an office to constitute being engaged in the business or regular activity of maintaining securities accounts within the meaning of Article 4(1)(a)(iii). The list includes being a place that is engaged solely in representational functions or administrative functions and lacks authority to make any binding decision to enter into any account agreement, unless the functions relate to the opening or maintenance of securities accounts. The list also includes *merely* being the place where the technology (*e.g.*, computers) supporting the bookkeeping or data processing is located; where a call centre is located; where mailing relating to securities accounts is organised; or where files or archives are located. None of the listed activities is a disqualification; the list serves only to set a bottom line – *any* of them *alone* would not constitute the office as being engaged in maintaining securities accounts within the meaning of Article 4(1)(a)(iii). It is clear that Article 4(2) does not apply in relation to Article 4(1)(a)(i) or (ii), compliance with which provides the parties with a safe harbour in satisfying the Qualifying Office test.

4-33 Where Article 4(1)(a)(i) or (ii) do not apply, the office will not constitute a Qualifying Office under Article 4(1)(a)(iii) *merely* by carrying on any *one* of the activities listed in Article 4(2). In any event, it may satisfy the Qualifying Office requirement if it falls under Article 4(1)(b) as an office identified by an account number, bank code, or other specific means of identification as maintaining securities accounts in the State whose law is chosen.

B. Detail of the provisions

1. Article 4(1)(a)

4-34 To be a Qualifying Office within Article 4(1)(a), an office needs to serve only *one* of the three functions listed. Moreover, that office is only required to be involved in *some* aspect of the particular function. Article 4(1) was drafted to take into account the fact that in today's market environment, different aspects of the various listed functions may be dispersed over several offices, or outsourced to third parties, in different locations (see paras. 4-24 *et seq.*). It is not necessary for these other offices or third parties to be located within the designated State.

4-35 The first of the three functions designated in sub-paragraph (i) of Article 4(1)(a) is the effecting or monitoring of entries to securities accounts.

4-36 The second listed function is the administration of payments or corporate actions relating to securities held with the intermediary. "Corporate actions" is a term used to describe various acts of the issuer in which the intermediary is involved on behalf of the account holder, such as the payment of cash dividends and other cash distributions, rights issues, bonus issues, reorganisation of share capital, forwarding proxies and performing other functions relating to the exercise of voting rights, reinvestment of dividends, transfers of securities, and the like.

4-37 The third listed function, which is relevant only if neither of the first two functions is exercised, is the residual category, "otherwise engaged in a business or regular activity of maintaining accounts." Nonexclusive examples of this residual category are the conclusion of account agreements relating to accounts which are to be held elsewhere; the provision of advisory and execution services relating to such accounts; and the receipt of instructions for acquisitions and dispositions of securities to be credited or debited to such accounts. Article 4(1)(a)(iii) is intentionally broad and designed to keep the list open to allow for new indicia of the activity of maintaining securities accounts.

2. Article 4(1)(b)

4-38 Article 4(1)(b) provides an alternative method of satisfying the Qualifying Office test. Without any inquiry into the functions that the relevant intermediary serves through that office (or any other), an office is a Qualifying Office if it is identified – by any specific means of identification (including an account number or a bank code) – as maintaining securities accounts in the agreed State. Neither an account number nor a bank code is a required, or even preferred, means of identification. Any specific means of identifying the office is sufficient. The Convention does not limit the mode of establishing a means of identification, nor does it limit the establishment of the means to the relevant intermediary; for example, a regulatory agency might establish such a means of identification. And, like the alternative method relating to functions served, this alternative does not relate in any way to the specific securities account or the specific securities.

3. Article 4(2)

4-39 Article 4(2) lists three classes of activity of an office none of which by itself is sufficient to constitute it a Qualifying Office under Article 4(1)(a)(iii). The first of these, technology supporting the record-keeping, has already been mentioned. The second is location or operation in the office of a call centre for communication with account holders. The third is the function of the office as a place where the mailing relating to securities accounts is organised or files or archives are located. In addition, an office is not a Qualifying Office where it is engaged in purely representative or administrative functions, unrelated to the opening or maintenance of securities accounts and has no authority to make any binding decision to enter into an account agreement. A representative office is an office of a foreign corporation or other legal entity which is designed to promote the corporation or entity, as by marketing and the provision of information, but has neither an independent legal existence nor (in contrast to a branch) any power to conclude agreements itself or make binding decisions to enter into such agreements. A representative office is not as such a Qualifying Office; nor is an office which performs purely administrative functions unrelated to the opening or maintenance of securities accounts, for example, personnel management.

4-40 While the pursuit of any or all of these activities does not *by itself* make an office a Qualifying Office, none is a disqualification and each may be an element which, when added to other factors, suffices to constitute the office a Qualifying Office when otherwise its status as such might be doubtful.

IV. Article 4(3)

4-41 The effect of Article 4(3) is that the definition of "relevant intermediary" applies not only to the holding of the account and to dispositions of securities to third parties but also to a disposition by an account holder to the intermediary itself, and this is so whether or not the intermediary maintains a securities account on its own records for which it is the account holder. The purpose of Article 4(3) is to make it clear, first, that although the intermediary in whose favour a transfer is made is in effect making a disposition to itself it is its customer who is the relevant account holder, and, secondly, to require that insofar as the transfer is matched by entries in the records of the intermediary's own intermediary, these are to be ignored in determining who is the relevant intermediary and what is the relevant account. Take the case where the intermediary does not hold an account for itself in its own books but maintains separate accounts for its own securities and for customers' securities in the books of its own (higher-tier) intermediary (a requirement in some legal systems in order to protect the account holder) and takes security from one of its account holders by effecting a transfer from its omnibus account with the higher-tier intermediary to its proprietary (or house) account. Then although all the credits are made through the accounts of the higher-tier intermediary it is the lower-tier intermediary that is the relevant intermediary for the purposes of the Convention; the account agreement between itself and the account holder from whom it has taken security is the relevant account agreement; and the securities account for the purposes of the second and third fall-back rules in Article 5 is the account maintained by the lower-tier intermediary as it stood immediately before the pledge or title transfer. This is logical because what is being pledged or transferred is the account holder's entitlement with its own (lower-tier) intermediary, not the account held in the books of the higher-tier intermediary. But for Article 4(3) the fact that the disposition was recorded in an account maintained by the upper-tier intermediary might have led to the conclusion that this was the relevant intermediary and the account it maintained was the relevant account.

Example 4-9

> An investor has securities credited to a securities account maintained for it by British Bank at its offices in London. The account agreement expressly provides that it is governed by the laws of England and does not expressly provide that a different law governs all the Article 2(1) issues. British Bank holds corresponding positions in the securities in an omnibus account with French Depository, based in Paris, under an account agreement expressed to be governed by French law. Investor transfers its securities to British Bank to secure a margin loan. No accounting entries are made in the records of French Depository. Later, in a proceeding in a State where the Convention is in force, the issue is whether the transfer was perfected. Under Article 4(3)(a), for purposes of applying Article 4(1) British Bank is the relevant intermediary and the relevant account agreement is the agreement between Investor and British Bank. The fact that the latter held a position with French Depository is irrelevant. Accordingly under Article 4(1) English law governs all the issues specified in Article 2(1), including the perfection issue.

Example 4-10

> The facts are as in the preceding illustration except that British Bank does not maintain an "own funds" account to record dispositions in its favour by its own account holders. Instead, it effects the transaction by procuring the transfer of the securities from its omnibus customer account with French Depository to its proprietary account with French Depository. Later, in a proceeding in a State where the Convention is in force, the issue before the court is whether the disposition by the account holder in favour of its own intermediary was perfected. Despite the fact that the title transfer is reflected only in the accounts maintained by French Depository on behalf of British Bank, it is the latter who is the relevant

intermediary, Investor who is the relevant account holder and its account with British Bank the relevant account. Accordingly, as in Example 4-9, the perfection requirement is governed by English law.

Example 4-11

Investor A has securities credited to a securities account maintained for it by Intermediary I. Investor A transfers the securities to Intermediary I as security for advances. The transfer is initially recorded solely in Intermediary I's own books but is later reflected in a transfer from Intermediary I's omnibus client account with its own intermediary to its proprietary account with that intermediary. This later transfer does not result in Intermediary I ceasing to be the relevant intermediary or the account agreement between Investor A and Intermediary I ceasing to be the relevant account agreement.

4-42 The phrase "immediately before the disposition" is to ensure that the securities account referred to when applying the fall-back rule of Article 5(2) or (3) is the account to which the securities had been credited immediately before the disposition rather than the account to which they were credited by reason of the disposition.

V. Dispositions by transfer from one securities account to another, including through a chain of intermediaries

4-43 Articles 4 and 5 apply independently with respect to each securities account (in other words, to each relationship between an account holder and its relevant intermediary). This principle is clear from the plain language of Articles 2, 4 and 5 and the definitions of the terms "securities held with an intermediary" and "relevant intermediary" (see commentary to Art. 1(1)(g) in paras. 1-17 and 1-18). It is further reinforced by Article 6(d), which provides that in determining the applicable law under the Convention, no account shall be taken of the place where any intermediary other than the relevant intermediary is located. Thus, when a chain of intermediaries stands between an account holder and the issuer, there is no single law that would govern all the Article 2(1) issues with respect to all securities accounts maintained by intermediaries standing between the account holder and the issuer. Similarly, where there is a disposition of securities by transfer from one securities account to another, including a disposition made through a chain of intermediaries, the independent application of Articles 4 and 5 with respect to each securities account may and often will result in a different law governing the Article 2(1) issues with respect to each account. The Convention rejects the idea of a "Super-PRIMA" which would apply the same law with respect to every securities account with each intermediary standing between an account holder and the issuer or with respect to every securities account involved in a disposition of securities made by transfer through a chain of intermediaries. Moreover, even in the simple case where there is a disposition of securities by transfer from one securities account to another maintained by the same intermediary, it is possible (though it rarely happens in practice) for the respective account agreements to be governed by different laws and therefore to result in one law governing the Article 2(1) issues with respect to the first account and another law governing the Article 2(1) issues with respect to the second account.

4-44 The examples set forth below illustrate some of the situations in which the independent application of the Convention with respect to each securities account can result in different laws governing the respective rights of different sets of people. It is assumed that the Qualifying Office test is satisfied in relation to each account agreement. The examples are examined and commented in paragraphs 4-45 *et seq.*

Example 4-12

A makes a disposition of securities to B, who immediately transfers them to C but fails to pay A. The dispositions are effected by a series of debits and credits to securities accounts maintained by a chain of intermediaries between A and B and another chain between B and C, including a debit to a securities account maintained for A by its intermediary and a credit to a securities account maintained for C by its intermediary (Y). A's account agreement is expressly governed by Ruritanian law, C's account agreement is expressly governed by Utopian law and Y's account agreement with its intermediary is governed by Sylvanian law. None of the account agreements specifies a different law to govern all the Article 2(1) issues. Suppose that under the Convention law determined by Article 4(1) with respect to C's account (Utopian law), C takes the securities credited to C's account free of all (incl. A's) adverse claims whereas under the Convention law determined with respect to A's account (Ruritanian law) the debit to A's account is subject to reversal because B failed to pay for the securities.

Example 4-13

The facts are as in Example 4-12, except that under the Convention law determined by Article 4(1) with respect to C's account (Utopian law), C does not acquire any rights in the securities greater than the rights that Y (or A, B or any intermediary other than Y in the chain between A and C) had to convey under the Convention law determined with respect to Y's (or such other person's) account (in the case of Y, Sylvanian law).

Example 4-14

The facts are as in Example 4-12, except that although the securities were credited to Y's account, Y did not credit them to C's account.

Example 4-15

A makes an improper double disposition by first pledging to B all the securities credited to A's securities account – using a method that does not give B exclusive control over A's securities account – and by later transferring the same securities to C. The transfer is effected by a series of debits and credits to securities accounts maintained by a chain of intermediaries between A and C, including a debit to a securities account maintained for A by its intermediary (X) and a credit to a securities account maintained for C by its intermediary (Y). A's account agreement is expressly governed by Ruritanian law, C's account agreement is expressly governed by Utopian law and Y's account agreement with its intermediary is expressly governed by Sylvanian law. None of the account agreements specifies a different law to govern all the Article 2(1) issues. Suppose that under the Convention law determined with respect to C's account (Utopian law), C takes free of B's interest, whereas under the Convention law determined with respect to A's account (Ruritanian law) the debit to A's account is subject to reversal because of B's interest or the transfer from A to C resulted in C's taking subject to B's interest.

Example 4-16

The facts are as in Example 4-15, except that under the Convention law determined with respect to C's account (Utopian law), C does not acquire any rights in the securities greater than the rights that Y (or A or any intermediary other than Y in the chain between A and C) had to convey under the Convention law determined with respect to Y's (or such other person's) account (in the case of Y, Sylvanian law).

Example 4-17

The facts are as in Example 4-15, except that although the securities were credited to Y's account, Y did not credit them to C's account.

Example 4-18

A holds an interest in 100 Z Corp shares credited to a securities account maintained for A by its intermediary (X). X, without instructions from A to do so, fraudulently debits 100 Z Corp shares to A's account and uses them to help make a disposition of 200 Z Corp shares to B, who takes in good faith without knowledge of X's fraudulent conduct. The disposition is effected by a series of debits and credits to securities accounts maintained by a chain of intermediaries between A and B, including the fraudulent debit of 100 Z Corp shares to A's account and a credit of 200 Z Corp shares to a securities account maintained for B by its intermediary (Y). A's account agreement is expressly governed by Ruritanian law and B's account agreement is expressly governed by Utopian law. Neither of the account agreements specifies a different law to govern all the Article 2(1) issues. Suppose that under the Convention law determined by Article 4(1) with respect to B's account (Utopian law), B takes all 200 Z Corp shares free of all (incl. A's) adverse claims, whereas under the Convention law determined with respect to A's account (Ruritanian law) the debit to A's account is subject to reversal because of X's fraud.

4-45 Each of these examples raises questions as to which law governs the respective rights of different sets of people. These are questions for which it was considered unnecessary to lay down a special Convention rule because the proper application of the general rule provides the correct solution, that is, the independent application of the Convention with respect to each securities account means that a person's rights resulting from a credit of securities to a particular securities account are governed by the Convention law determined with respect to that account. This general rule applies independently of whether the account holder's rights are enforceable against its intermediary or directly against the issuer (see further comments in para. 4-50). So the question in Example 4-12 – whether C takes free of all (incl. A's) adverse claims, once the securities were credited to C's account – is answered by the Convention law determined with respect to that account, namely Utopian law. Because Ruritanian law, as the Convention law determined with respect to A's account, treats the debit of securities to that account as subject to reversal, A may, under Ruritanian law, continue to have rights against its intermediary; these rights can be proprietary, contractual, mixed or other in character. But this effect under Ruritanian law does not affect whether C takes the securities credited to its account free of all (incl. A's) adverse claims; that issue is governed exclusively by the Convention law determined with respect to C's account. The substantive content of that (Utopian) law may provide that C takes free of all adverse claims (as in Example 4-12) or that C acquires only the rights that Y (or A, B or any intermediary other than Y in the chain between A and C) had to convey under the Convention law determined with respect to Y's (or such other person's) account (as in Example 4-13). But in both cases that issue is governed exclusively by the Convention law determined with respect to C's account, the account to which those securities were credited for C's benefit. In Example 4-14, where the securities were debited from A's account and transferred by a series of debits and credits to securities accounts maintained by a chain of intermediaries between A and C, including a credit to Y's account, but were not credited to C's account, the question whether C has any rights resulting from the credit of securities to Y's account is governed by the Convention law determined with respect to Y's account. Even if C did not become the owner under that law (Sylvanian law), however, that law might give C the right to compel Y to credit the securities to C's account based on the contractual terms of its account agreement with Y. Also, C might find an independent source of rights against Y under the Convention law with respect to C's account (Utopian law), that is, Utopian law might give C rights whether or not Y has actually credited the securities to C's account and/or might give C the right to compel Y to credit the securities to C's account based on the contractual terms of its account agreement with Y. In addition, C might be given rights under any other law binding on Y.

4-46 Similarly, the questions in Examples 4-15 and 4-16 – whether C takes the securities subject to B's security interest, acquires priority over B's security interest, or takes free of all adverse claims – is answered by the Convention law determined with respect to C's account (Utopian law). In Example 4-15, C takes free of B's interest. Although B may continue to have rights against A's intermediary and the intermediary's property because the Convention law determined with respect to A's account (Ruritanian law) treats the debit of securities to A's account as subject to reversal and treats B's security interest as still effective, this does not affect whether C takes subject to B's security interest, acquires priority over B's security interest, or takes free of all (incl. A's) adverse claims – that issue is governed exclusively by the Convention law determined with respect to C's account (Utopian law). Thus, it is not possible to have a situation in which C acquires an interest in securities credited to C's account that, under Utopian law, is free of all adverse claims, but at the same time to be vulnerable to the successful assertion by B or any other person of rights against C under Utopian or any other law. The applicable law may provide that C takes free of all adverse claims (as in Example 4-15) or that C acquires only the rights that Y (or A or any intermediary other than Y in the chain between A and C) had to convey under the Convention law determined with respect to Y's (or such other person's) account (as in Example 4-16). But in both cases that issue is governed exclusively by the Convention law determined with respect to C's account, the account to which those securities were credited for C's benefit. In Example 4-17, the analysis is the same as the analysis that is discussed for Example 4-14 in para. 4-45 above.

4-47 Similarly, the questions in Example 4-18 – whether B takes all 200 Z Corp shares credited to its account free of all (incl. A's) adverse claims, once the Z Corp shares were credited to B's account – is answered by the Convention law determined with respect to that account, namely Utopian law. Because Ruritanian law, as the Convention law determined with respect to A's account, treats the debit of 100 shares of Z Corp shares to that account as subject to reversal, A may, under Ruritanian law, continue to have rights against X and its property that are proprietary, contractual, mixed or other in nature. But this treatment under Ruritanian law does not affect whether B takes all 200 Z Corp shares free of all (incl. A's) adverse claims; that issue is governed exclusively by the Convention law determined with respect to B's account. This is true despite the fact that under Ruritanian law, X's act was fraudulent. Although A's ability to discover B's existence and to assert a claim based on the argument that "its" 100 shares ended up in B's account is extremely unlikely to occur given the nature of the intermediated system, this example demonstrates the application of the Convention's rule even in such situations.

4-48 In the simple (but unusual) case where the accounts maintained for A and C are maintained by the same intermediary but the account agreements are governed by the different laws indicated (Variants of the patterns in Examples 4-12 and 4-15), the intermediary would be required by Utopian law to honour the credit of securities to C's account while being required by Ruritanian law to reverse the debit to A's account. In the absence of any reallocation of this risk in either of the two account agreements, the intermediary would likely be required by applicable regulatory or property law, or as a matter of contractual obligation, to acquire additional securities on the open market to ensure that it had enough securities to match the credits of securities to both A's account and C's account or to give one of the account holders the cash value of the securities in lieu of the securities themselves. In the more complex (but realistic case) where the two accounts are governed by different laws but maintained by separate intermediaries and there is a chain of intermediaries between A's account and C's account, either A's intermediary or C's intermediary, or some other intermediary standing between A and C, would suffer a double liability (in the sense that its transfer to C would not relieve it of a continuing liability to A) in the absence of provisions in its account agreements with the relevant account holder reallocating the risk of such double liability to the account holder or spreading it to all of the intermediary's account holders.

4-49 It is important to note that this risk of double liability for intermediaries has *always* existed with respect to cross-border transfers of securities through a chain of intermediaries, is well-understood by informed intermediaries involved in cross-border transfers, and was *not* created by the Convention. Under existing conflict of laws rules, different laws can and often do govern issues similar to those listed in Article 2(1) with respect to securities issued by issuers organized under the laws of a variety of States but credited to a single securities account or disposed of through securities accounts maintained by a chain of intermediaries in a variety of States. Thus, informed intermediaries involved in cross-border transfers of securities typically reallocate this risk in their account agreements with specific account holders, spread the risk over a pool of account holders, or consciously decide to assume and manage this risk. The Convention does not attempt to eliminate this risk, but it does make it easier for intermediaries to identify, reallocate or manage this risk by making it clear which State's laws govern all the Article 2(1) issues with respect to each securities account.

4-50 It is also important to emphasize that the analysis is the same in the case in which the Convention law determined with respect to securities accounts maintained by an intermediary treats the intermediary as a mere record keeper and treats the credit of securities to those accounts as establishing a direct relationship between the account holder and the issuer (or the intermediary's intermediary). It is the Convention law determined with respect to a transferee's account that governs whether the transferee takes an interest in securities credited to its securities account free of all adverse claims. The fact that, under the substantive content of that (or any other) law, the transferee is deemed to hold direct from the issuer is irrelevant to the application of the Convention (see paras. Int-19 *et seq.*).

4-51 As illustrated in Example 4-13, if, under the Convention law determined with respect to C's account, (Utopian law), C cannot acquire an interest in the securities greater than the rights that Y (or A, B or any intermediary other than Y in the chain between A and C) had to convey, the priority issue is nevertheless governed by Utopian law. If Utopian law looks to the Convention law determined with respect to Y's (or such other person's) account to determine what rights Y (or such other person) had to convey and thus C acquired, it does so not because the Convention says that the Convention law determined with respect to Y's (or such other person's) account governs the priority issue, but because the substantive content of the Convention law determined with respect to C's account (Utopian law) effectively says that Sylvanian (or such other) law determines what rights Y (or such other person) had to convey.

Article 5 *Fall-back rules*

1. If the applicable law is not determined under Article 4, but it is expressly and unambiguously stated in a written account agreement that the relevant intermediary entered into the account agreement through a particular office, the law applicable to all the issues specified in Article 2(1) is the law in force in the State, or the territorial unit of a Multi-unit State, in which that office was then located, provided that such office then satisfied the condition specified in the second sentence of Article 4(1). In determining whether an account agreement expressly and unambiguously states that the relevant intermediary entered into the account agreement through a particular office, none of the following shall be considered –

a) a provision that notices or other documents shall or may be served on the relevant intermediary at that office;

b) a provision that legal proceedings shall or may be instituted against the relevant intermediary in a particular State or in a particular territorial unit of a Multi-unit State;

c) a provision that any statement or other document shall or may be provided by the relevant intermediary from that office;

d) a provision that any service shall or may be provided by the relevant intermediary from that office;

e) a provision that any operation or function shall or may be carried on or performed by the relevant intermediary at that office.

2. If the applicable law is not determined under paragraph (1), that law is the law in force in the State, or the territorial unit of a Multi-unit State, under whose law the relevant intermediary is incorporated or otherwise organised at the time the written account agreement is entered into or, if there is no such agreement, at the time the securities account was opened; if, however, the relevant intermediary is incorporated or otherwise organised under the law of a Multi-unit State and not that of one of its territorial units, the applicable law is the law in force in the territorial unit of that Multi-unit State in which the relevant intermediary has its place of business, or, if the relevant intermediary has more than one place of business, its principal place of business, at the time the written account agreement is entered into or, if there is no such agreement, at the time the securities account was opened.

3. If the applicable law is not determined under either paragraph (1) or paragraph (2), that law is the law in force in the State, or the territorial unit of a Multi-unit State, in which the relevant intermediary has its place of business, or, if the relevant intermediary has more than one place of business, its principal place of business, at the time the written account agreement is entered into or, if there is no such agreement, at the time the securities account was opened.

I. Introduction

5-1 Article 5 sets out a series of "fall-back" rules that apply if the law governing the Article 2(1) issues is not determined under the Convention's primary rule (Art. 4(1)). This might occur if an account holder and the relevant intermediary have not expressly agreed on a law governing their account agreement or on another law governing all the Article 2(1) issues; or, if there is such an agreement, the designation of the governing law is not given effect by the Convention because the Qualifying Office requirement is not satisfied. The fall-back rules in Article 5 operate as a cascade. Thus, if the applicable law is not determined under Article 4, the first fall-back rule is Article 5(1); if the applicable law is not determined under Article 5(1), the next fall-back rule is Article 5(2); and if the applicable law is not determined under Article 5(2), Article 5(3) applies.

5-2 Application of the Convention's conflict rules (including Art. 5) in the context of a disposition of securities effectuated by transfer from one securities account to another, including through a chain of intermediaries, is discussed at length and illustrated by useful examples in paragraphs 4-43 *et seq.*

5-3 Under Article 12(3), a Multi-Unit State may make a declaration to the effect that if the applicable law, determined under any of the fall-back rules in Article 5, is that of the declaring Multi-unit State itself or that of one of its territorial units, the *internal* conflict of laws rules in force in that Multi-unit State must be applied (on this form of *internal renvoi* within Multi-unit States in the context of Art. 5, see paras. 12-16 *et seq.*).

II. The first fall-back rule: Article 5(1)

5-4 If a *written* account agreement *expressly and unambiguously states* that the relevant intermediary entered into the agreement through a particular office, the law in force in the State, or the territorial unit of a Multi-unit State, in which that office was located at the time of entry into the account agreement is the law applicable to all the issues specified in Article 2(1), provided that that office *then* satisfied the condition specified in the second sentence of Article 4(1) (*i.e.*, was a Qualifying Office). On the meaning of the expression "law in force", see the comments in paragraphs 4-15, 12-12 and 12-13.

5-5 The first fall-back rule operates so as to apply the law of an office's location only where:

(1) there is a written account agreement;

(2) the agreement expressly and unambiguously states that it was entered into by the relevant intermediary through the office in question; and

(3) the designated office is a Qualifying Office.

"Written" is defined in Article 1(1)(n) (see para. 1-29). What constitutes a Qualifying Office has already been discussed (see paras. 4-21 *et seq.*). That leaves for consideration the requirement of an express and unambiguous statement in the account agreement that it was entered into through a particular office.

5-6 In any event, the first fall-back rule is based exclusively on the wording of the account agreement. This is to promote certainty and to avoid disputes. Implication or an ambiguous statement does not suffice. To emphasize this point Article 5(1) lists five factors to be excluded in determining whether the agreement contains the requisite express and unambiguous statement. These are provisions as to:

(a) the office at which notices or other documents shall or may be served on the relevant intermediary;

(b) the State, or territorial unit of a State, in which proceedings shall or may be instituted against the relevant intermediary;

(c) the office from which a statement or other document shall or may be provided by the relevant intermediary;

(d) the office from which any service shall or may be provided by the relevant intermediary;

(e) the office at which any operation or function shall or may be carried on or performed by the relevant intermediary.

So the inclusion of any or all of these provisions does not suffice as an express and unambiguous statement of the office through which the agreement was entered into.

Example 5-1

An intermediary organised under the laws of Italy and its customer enter into an account agreement. The account agreement does not include a choice of law clause, and it does not expressly provide that a particular law governs all the Article 2(1) issues. It does expressly and unambiguously (in a separate paragraph on the signature page of the account agreement) state that the intermediary entered into the account agreement through its Frankfurt office. The identified Frankfurt office was a Qualifying Office at the time the account agreement was entered into. Later, the customer grants to a lender a security interest in securities credited to the account which were issued by issuers incorporated in Russia, Poland and Luxembourg. Still later, in a proceeding in a State where the Convention is in force, the issue is whether the security interest is perfected. Under Article 5(1), the law of Germany governs all the Article 2(1) issues, including the issue of perfection. The fact that the intermediary was organised under Italian law and the fact that the issuers were all incorporated in countries other than Germany are irrelevant.

III. The second fall-back rule: Article 5(2)

5-7 The second fall-back rule looks to the State, or the territorial unit of a Multi-unit State, under whose law the relevant intermediary was incorporated or otherwise organised at the time the written account agreement was entered into or, if there is no written account agreement or no account agreement at all, at the time the securities account was opened. In the great majority of cases a securities account will be opened pursuant to a prior account agreement, even if initially only informal. But this will not always be the case. In some circumstances the rights of the account holder and the intermediary are fixed not by agreement but by law. For example, a Nordic CSD, which qualifies as an intermediary under

Article 1(4) even if not performing an intermediary's functions, may not always have account agreements with its customers in relation to securities issued in its own country but operates under legal rules which determine the rights and duties of the parties (see comments in para. 1-7). In such a case the relevant time is that of the opening of the securities account.

5-8 The second fall-back rule deals with two types of intermediary: one that is "incorporated" under a given law and one that is "otherwise organised" under a given law. A corporation is a body having a legal personality distinct from that of its members. An intermediary which is "otherwise organised" includes an unincorporated association which, though it may not have a distinct legal personality, nevertheless has a status in law derived from the fact that it is formed in accordance with specified legal rules and thus is more than simply a group of natural persons coming together in business. In some jurisdictions certain types of partnership would fall within this category. In summary, the second fall-back rule covers all forms of association recognised by law, incorporated and unincorporated and whether or not having a distinct legal personality.

5-9 The second fall-back rule is envisaged as coming into operation only in a very small percentage of cases. Its justification is that it provides predictability in a situation where Article 4(1) and Article 5(1) have not determined the applicable law.

5-10 The second fall-back rule also deals with the case where, in a Multi-unit State, the intermediary is incorporated or otherwise organised not under the law of a territorial unit but under the law of the State (*e.g.*, Canada has banks incorporated under federal law). In the case of a federal bank reference to the law of the place of incorporation would not by itself identify the province whose law was to apply. Article 5(2) applies the law in force in the province in which the bank has its place of business or, if more than one, its principal place of business.

IV. The third fall-back rule: Article 5(3)

5-11 Where neither of the preceding fall-back rules apply, the law to be applied is that in force in the State, or the territorial unit of a Multi-unit State, in which the relevant intermediary has its place of business, or if more than one such place, its principal place of business, at the time the written account agreement was entered into or, if there was no such agreement, at the time the securities account was opened. For purposes of the Convention (both Arts. 5(2) and (3)), the principal place of business is the place from which the intermediary's business is managed, that is, its head office or chief executive office (often also referred to as its place of central administration).

V. Amendment of the account agreement

5-12 In general, the applicable fall-back rule under Article 5 has to be determined by reference to the relevant factor (statement in written account agreement, place of incorporation or place of business) existing at the time the parties enter into the written account agreement, or, if there is no written account agreement or no account agreement at all, the securities account is opened. There may be circumstances in which an amendment of the account agreement may lead to a change of the applicable law (see comments in paras. 7-1 and 4-27 *et seq.*).

Article 6 ***Factors to be disregarded***

In determining the applicable law in accordance with this Convention, no account shall be taken of the following factors –

a) the place where the issuer of the securities is incorporated or otherwise organised or has its statutory seat or registered office, central administration or place or principal place of business;

b) the places where certificates representing or evidencing securities are located;

c) the place where a register of holders of securities maintained by or on behalf of the issuer of the securities is located; or

d) the place where any intermediary other than the relevant intermediary is located.

6-1 In order to prevent courts from applying to securities held with an intermediary conflict of laws rules devised for directly held securities, Article 6 provides that in determining the applicable law in accordance with the Convention no account is to be taken of the place where the issuer of the securities is incorporated or formed or has its statutory seat or registered office or place or principal place of business or of the place where certificates representing or evidencing securities are located or the place where a register of holders is maintained by or on behalf of the issuer. Moreover, in applying the rule in Article 16(4), no consideration may be given to any of these factors in determining whether there is an agreement as to the State in which the securities account is maintained implied from the terms of the account agreement considered as a whole or the surrounding circumstances. All these places provide a traditional connecting factor for directly held securities but they have no role to play in relation to indirect holdings, for which the Convention focuses exclusively on the relevant intermediary and the relationship between that intermediary and its account holders. Hence Article 6 also excludes from consideration the location of any intermediary other than the relevant intermediary.

6-2 Article 6, taken with Articles 4 and 5, reflects a key policy decision, namely the rejection of any conflict of laws approach that would involve looking through an account holder's intermediary to a higher-tier intermediary or to the issuer itself to determine the applicable law. So even in a jurisdiction under whose substantive law the account holder has the right to enforce the securities directly against the issuer or is otherwise treated as the direct owner of the securities, that fact would be irrelevant in determining the Convention law.

6-3 Although the factors mentioned in Article 6 are to be disregarded in determining the applicable law, they remain relevant to the question whether the situation under consideration involves an international element for the purpose of Article 3 (see comments in para. 3-3). Moreover, Article 6 has no relevance to the applicability of Article 16 in cases where the parties have agreed that the securities account is maintained in a particular State or territorial unit.

Article 7 ***Protection of rights on change of the applicable law***

1. This Article applies if an account agreement is amended so as to change the applicable law under this Convention.

2. In this Article –

a) "the new law" means the law applicable under this Convention after the change;

b) "the old law" means the law applicable under this Convention before the change.

3. Subject to paragraph (4), the new law governs all the issues specified in Article 2(1).

4. Except with respect to a person who has consented to a change of law, the old law continues to govern –

a) the existence of an interest in securities held with an intermediary arising before the change of law and the perfection of a disposition of those securities made before the change of law;

b) with respect to an interest in securities held with an intermediary arising before the change of law –

i) the legal nature and effects of such an interest against the relevant intermediary and any party to a disposition of those securities made before the change of law;

ii) the legal nature and effects of such an interest against a person who after the change of law attaches the securities;

iii) the determination of all the issues specified in Article 2(1) with respect to an insolvency administrator in an insolvency proceeding opened after the change of law;

c) priority as between parties whose interests arose before the change of law.

5. Paragraph (4)(c) does not preclude the application of the new law to the priority of an interest that arose under the old law but is perfected under the new law.

I. Introduction

7-1 Article 7 determines the impact of an amendment to an account agreement if the consequence of the amendment is that the Convention law changes from the law of one State or territorial unit of a Multi-unit State (reference to territorial units will not be repeated throughout this commentary to Art. 7) as determined under either Article 4(1) or Article 5, to the law of a different State, as determined under Article 4(1), which assumes that the Qualifying Office requirement of the second sentence of Article 4(1) is satisfied at the time of the amendment (a 'triggering amendment'). In the case where the parties fail to select a new applicable law under the Convention which satisfies the requirements of Article 4(1), the amendment is ignored for Convention purposes (*i.e.*, the *status quo ante* is preserved). The amendment does not lead to the application of Articles 5(2) or (3); and Article 5(1) cannot be triggered by the amendment because Article 5(1) can be applied only at the time the account agreement was originally entered into (*i.e.*, the initial creation of the securities account) and it does not constitute a triggering amendment for purposes of Article 7. Further, Article 7 does not apply when a different Convention law is brought into play as the result of a transfer of securities to a new account; Article 7 applies only if the change of the Convention law is the consequence of an amendment to an account agreement. This principle suggests that the sphere of application of Article 7 will likely be rather circumscribed in actual practice. As a consequence of this principle, Article 7 is not brought into play when a security interest is implemented by means of a transfer of securities from a collateral provider's account to a collateral taker's account (as the rights resulting from the credit of securities to the collateral taker's account, as well as all the other Article 2(1) issues, will be governed by the Convention law determined with respect to the collateral taker's account – the same, of course, is true with respect to securities transferred outright to a buyer). Therefore, references to the grant of a security interest in this commentary to Article 7 refer to a method of granting that does not result in a credit of the securities to the securities account of the collateral taker, a method which is, for instance, in current practice often used in Switzerland. Finally, an amendment is a triggering amendment only if it expressly (i) initiates or changes an agreement on governing law, or (ii) singles out and expressly reconfirms an original agreement on governing law that was made at a time when the Qualifying Office requirement was not satisfied; an amendment which changes other matters does not meet this criterion (*i.e.*, does not trigger a new application of the Qualifying Office requirement) even if that amendment takes the form of restating the entire agreement (see further comments in para. 4-27).

7-2 The following fact patterns illustrate when Article 7 applies:

(1) The original Convention law is determined by application of Article 4(1); the agreement on governing law is amended; the amendment results in a change of the Convention law by application of Article 4(1). Article 7 applies.

(2) The original Convention law is determined under Article 5; the account agreement is amended by the addition of an express agreement on governing law; the amendment results in a change of the Convention law by application of Article 4(1). Article 7 applies.

(3) The original Convention law is determined under Article 4 or 5; the account agreement is amended, but the amendment does not result in a change of the Convention law by application of Article 4(1). Article 7 does not apply.

To sum up, Article 7 applies only if a change of law results from a triggering amendment, which means that the new law is determined by application of Article 4(1).

7-3 Article 7 has no forerunner in pre-Diplomatic Conference drafts because the question of how to deal with existing rights in the event of a change of law by virtue of a triggering amendment arose only when the Diplomatic Conference decided to select as the primary conflict of laws rule the (albeit conditioned) choice of law approach in Article 4(1).

7-4 Article 7 deals with a change of law in consequence of a triggering amendment under the Convention, that is, Convention law A is replaced by Convention law B as a result of a triggering amendment. Thus, the change of law dealt with in Article 7 is one that occurs after the Convention entered into force. In contrast, Article 15 is a transition provision which provides that the Convention law determines the priority of competing interests in securities arising before and after the Convention enters into force in a Contracting State. Hence, Articles 7 and 15 deal with completely different factual situations. However, both articles are similar (only) insofar as the new law governs (albeit with exceptions in Art. 7(4)) the Article 2(1) issues.

7-5 Triggering amendments may occur at any time. The account holder and its intermediary are able to change the applicable law at any time without the consent of third parties. This generates the need to describe clearly the consequences of the change of law for third parties, who often will have no knowledge of the triggering amendment or may not be in a position to effectively protect themselves. Therefore, while under Article 7(3) the new law applies in general, Article 7(4) preserves the applicability of the old law in specified situations with respect to interests of specified parties.

7-6 Article 7(3) states the general rule – the law applicable under the Convention after a change of law (the 'new law') governs all the Article 2(1) issues in respect of any interest in securities previously or subsequently credited to the securities account governed by the amended account agreement. However, where all relevant events occurred before the change, the old law governs the issues specified in Article 7(4), unless asserted by a person who consented to the change (and subject to Art. 7(5)). A person who consents to the change does not fall within Article 7(4). Under Article 7(4), the old law continues to govern specified issues with respect to certain interests in securities credited to a securities account that were acquired before the triggering amendment by a person who did not consent to the triggering amendment ('protected person'). In favour of those protected persons, the old law continues to govern the existence of an interest that arose prior to the change and the perfection of a disposition made before the change (Art. 7(4)(a)). A similar rule is specified in Article 7(4)(b) for various situations: Sub-paragraph (b)(i) provides that the old law continues to govern the legal nature and effects of such an interest against the relevant intermediary and any party to a disposition of those securities made before the change of law. Sub-paragraphs (b)(ii) and (iii) provide that the old law continues to govern the status of an interest that arose before the change of law against attaching creditors and all the Article 2(1) issues with respect to insolvency administrators, when the attachment or the insolvency proceeding occurs after the change of law. Sub-paragraph (c) preserves the applicability of the old law to priority issues as between pre-change interests, except (Art. 7(5)) situations where at least one of those competing interests arose under the old law, was not perfected under the old law but subsequently is perfected under the new law.

7-7 In summary, Article 7(1) determines the scope of the provision. Article 7(2) defines key terms used in Article 7. Article 7(3) states the general rule – the law applicable under the Convention after a change of law (the 'new law') governs all the Article 2(1) issues in respect of any interest in securities previously or subsequently credited to the securities account governed by the amended account agreement, subject to the exceptions specified in Article 7(4). Article 7(4) provides, subject to Article 7(5), that the old law continues to govern certain specified issues with respect to an interest in securities credited to an account that was acquired before the change of law by a protected person, that is, a person other than a person who consented to the change of law. Article 7(5) provides an exception to the exception contained in Article 7(4)(c) in case of certain priority issues.

II. Article 7(1): Scope of the provision

7-8 Article 7 applies only if there is a triggering amendment (see paras. 7-1 and 7-2). A triggering amendment may add an Article 4(1) governing law clause where none existed previously. A triggering amendment may replace an already existing governing law clause that might or might not have had the effect of determining the old law under Article 4(1). If there was no previous governing law clause that had the effect of determining the Convention law under Article 4(1), then the applicable law will initially have been determined by Article 5, and the amendment will be a triggering amendment (which produces an applicable law determined under Art. 4(1)) if it results in the applicability of a law different from that previously determined under Article 5. As Article 16(1) states that references in the Convention to an account agreement include an account agreement concluded before this Convention entered into force on the international plane (see Art. 19(1)), Article 7 applies also to such pre-Convention account agreements (see Example 16-1 *in fine*).

7-9 Applicability of Article 7 presupposes that

(a) an account agreement exists before any amendment is made;

(b) the actions taken by the parties constitute an amendment to the account agreement governing the securities account and not the entering into of a new account agreement governing a different securities account;

(c) the amendment causes a change of law by reason of the application of Article 4(1) (triggering amendment; see Example 7-1 below).

If all these elements are present, Article 7 applies whether the applicable law determined under the Convention *before* the triggering amendment is made was determined under Article 4(1) (whether or not with the assistance of the interpretive rules of Art. 16) or under Article 5. If any of these elements is not present, Article 7 does not apply.

7-10 Article 7 does not affect situations where the State whose law applies under the Convention remains the same but the content of that law changes (see Example 7-2). Those situations are addressed by the transition rules in that State's law.

7-11 If the parties entered into a new account agreement, rather than amending an existing agreement, this constitutes the creation of a different securities account and, of course, Article 7 will have no application; the law governing all the Article 2(1) issues with respect to securities credited to the new account will be determined under Article 4 or 5, as applicable. This is not a change of the Convention law determined with respect to a particular account, and Article 7 is not applicable.

7-12 Article 7 is likewise not applicable, when the law changes as a consequence of a transfer of securities from one securities account to another, each of which is governed by a separate account agreement (see Example 7-3, as well as the comments in paras. 7-1, 4-11 and 4-43 *et seq.*).

7-13 Nor does Article 7 apply to a situation if the original Convention law was determined under Article 5(2) or (3) and the relevant intermediary's State of incorporation, organisation, place of business or principal place of business changes. In this case there is not even a change of law, because a subsequent change of one of these connecting factors does not change the law under Article 5(2) or (3).

Example 7-1

An intermediary organised under the laws of Jordan and its customer enter into an account agreement. The account agreement expressly provides that all the Article 2(1) issues are governed by English law. The intermediary had a Qualifying Office in London at the time the governing law agreement was made. Later, the intermediary and its customer amend the account agreement so that it expressly provides that all the Article 2(1) issues are governed by German law. However, the intermediary does not have a Qualifying Office in Germany at the time of the amendment. Because the Qualifying Office requirement (Art. 4(1), second sentence) is not satisfied, there is no change of law; the amendment (whatever effect it may have in other respects) is not a triggering amendment. Article 7 does not apply. English law continues to govern all the Article 2(1) issues.

Variant:

If the intermediary had a Qualifying Office in Germany at the time of the amendment, it was a triggering amendment and Article 7 applies.

Example 7-2

An intermediary enters into an account agreement with its customer. Assume that the Convention determines that Ruritanian law is the Convention law. The account holder grants to secured party A (SP-A) a security interest in the securities account and all the securities credited to the account. SP-A perfects the security interest in accordance with Ruritanian law. Later, Ruritania amends its law to change the steps that are required to perfect a security interest. Under the transition rules of the new Ruritanian law, secured parties with perfected security interests must comply with the perfection procedures of the new Ruritanian law within six months of its enactment in order to remain perfected. SP-A fails to do so. Still later, the account holder grants a security interest in the securities account and all the securities credited to the account to SP-B, which perfects its security interest under the new Ruritanian law. Article 7 does not apply to this situation because the change in legal rules was neither a change of law within the meaning of Article 7 nor a consequence of a triggering amendment by the parties. Ruritanian law was not replaced as the applicable law by the law of a different State (or different territorial unit of a Multi-unit State); there was merely a change in the content of

Ruritanian law. Such a change does not trigger Article 7 and the effect of such a change is not determined by the Convention.

Example 7-3

An intermediary organised under the laws of Germany, with Qualifying offices in Ruritania and Utopia, enters into two separate account agreements with a customer. One account agreement expressly and unambiguously provides that the intermediary is entering into the agreement through its Ruritanian office (the "Alpha account agreement") and the other so provides through its Utopian office (the "Beta account agreement"). Neither of the account agreements expressly provides that a particular law governs the agreement or all the Article 2(1) issues. Each of the account agreements governs a separate securities account (the "Alpha account", and the "Beta account" respectively). The account holder grants a security interest in the Alpha account and all securities credited to that account to a bank organised under English law. The collateral taking English bank perfects the security interest in accordance with Ruritanian law by a method that does not result in a credit of the securities to its securities account. Later, the account holder transfers all the securities from its Alpha account to its Beta account and subsequently grants a security interest in the Beta account and all securities credited to that account to a lender organised under French law. The collateral taking French lender perfects its security interest under Utopian law by a method that does not result in a credit of the securities to its securities account. The Convention law that governs all the Article 2(1) issues with respect to the French lender's rights is Utopian law. Article 7 does not apply to this situation because the applicable law is Utopian law, not by virtue of a triggering amendment to the account agreement, but as the consequence of the transfer of the securities from the "Alpha account" to the "Beta account".

III. Article 7(2): Definitions of 'old' and 'new' law

7-14 For drafting convenience, Article 7 uses the terms 'old law' and 'new law'. Article 7(2) defines these terms. The term 'old law' means the law of the State applicable under the Convention before a triggering amendment is made. The term 'new law' means the law of the other State made applicable under the Convention as a consequence of a triggering amendment. In the unlikely case of several successive triggering amendments, each of these amendments "creates" an "old" and a "new" law. However, the protective rules of Article 7(4) and (5) apply even against a "later" new law. Thus, in the case of two successive triggering amendments, as a result of which laws A, B and C successively apply, a security interest perfected under law A is protected by Article 7(4), for example in insolvency proceedings, not only vis-à-vis the change to law B but also the later change to law C.

IV. Article 7(3) – General rule: Applicability of the 'new' law

7-15 Article 7(3) states the general rule: the 'new law' governs all the issues specified in Article 2(1) unless one of the exceptions in paragraph 4 applies.

Example 7-4

An intermediary organised under the laws of Luxembourg and its customer enter into an account agreement. The account agreement expressly provides that all the issues specified in Article 2(1) are governed by Ruritanian law. The intermediary had a Qualifying Office in Ruritania at the time the governing law agreement was made. The account holder grants a security interest in the securities account and all the securities credited to the account to a bank organised under English law. The collateral taking bank perfects the security interest in accordance with Ruritanian law by a method that does not result in a credit of the securities to its securities account. Later, the intermediary and its customer amend the account agreement so that it expressly provides that all the issues specified in Article 2(1) are governed by Utopian law. The intermediary had a Qualifying Office in Utopia at the time of the amendment. The account holder grants a security interest in the securities account and all the securities credited to the account to a lender organised under French law. The collateral taking lender perfects the security interest in accordance with Utopian law by a method that does not result in a credit of the securities to its securities account. Later, in a proceeding in a State where the Convention is in force, the issue is whether the security interest of the bank has priority over the security interest of the lender. Article 7 applies to this situation because the Luxembourg intermediary and its customer amended the governing law provision of the account agreement, the consequence of that amendment was that the applicable law determined by the Convention to govern all the Article 2(1) issues changed from Ruritanian law to Utopian law, and the new law (Utopian law) was determined under Article 4(1). Pursuant to the general rule in Article 7(3), the new law (Utopian law) determines the priority of these security interests; none of the exceptions in Article 7(4) applies (note that in this example, one of the security interests arose after the change of law). The fact that the new law (Utopian law) governs the priority issue does not necessarily mean that the security interest of the lender will have priority over the security interest of the bank. The outcome of that question depends on the content of the new substantive law (Utopian law).

V. Article 7(4) – Exceptions: Applicability of the 'old' law

7-16 Article 7(4) provides that the old law continues to govern the *following issues* with respect to an interest in securities credited to a securities account and which was *acquired before* the change of law by a *protected person*, that is, a person other than a person who consented to the change of law:

(a) whether the interest exists (Art. 7(4)(a));

(b) whether the person who acquired the interest completed the steps necessary to render the disposition by which it acquired the interest effective against persons who are not parties to that disposition (Art. 7(4)(a));

(c) the legal nature and effects of the interest against the relevant intermediary and any party to a disposition of those securities made *before* the change of law (Art. 7(4)(b)(i));

(d) the legal nature and effects of the interest against a person who attaches the securities *after* the change of law (Art. 7(4)(b)(ii)); clearly, as against an attachment occurring *before* the change of law, the old law applies;

(e) all the Article 2(1) issues with respect to an insolvency administrator in an insolvency proceeding opened *after* the change of law (Art. 7(4)(b)(iii)); the restriction to issues "with respect to the insolvency administrator" avoids overlapping with issues dealt with in earlier sub-paragraphs; and

(f) the priority of competing interests, any one or more of which arose before the change of law, subject to the possible displacement of that priority by application of the new law to the priority of an interest that arose under the old law, was not perfected under the old law but subsequently is perfected under the new law (Art. 7(5)).

7-17 Article 7(4) does not apply to dispositions made *after* a change of law. Thus, it does *not* preserve the old law to govern whether an interest in securities acquired by a person *before* a change of law has priority over the interest of another person acquired by a disposition made *after* the change of law, even if the person receiving the disposition is a person who consented to the change of law. The matter is to be decided by the new law (Art. 7(3)).

7-18 The opening words of Article 7(4) make clear that the rules in its sub-paragraphs do *not* preserve the old law to govern any of the issues specified therein with respect to any interest in securities acquired before a change of law was made by a person who *consented* to that change of law. In addition to the parties to an account agreement, third parties too may consent to a particular change of law (*e.g.*, a secured party whose consent to the change of law was obtained by the account holder or the relevant intermediary before or after the change of law). Consent eliminates any need for protection. Thus, the holder of a security interest perfected without registration under the old law, may, by consenting to a change of law, run the risk that the interest will not be perfected after the change of law for want of registration or that the change of law will result in a loss of rank with respect to other secured parties who acquired their secured interests before the change of law. With respect to the interest of any consenting party, the general rule of Article 7(3) applies (*i.e.*, the new law governs all the Art. 2(1) issues), except to the extent Article 7(4) applies when the competing party's interest is that of a protected person. If Article 7(4) applies with respect to that competing interest, the old law governs all the issues specified in the relevant category of Article 7(4), even when it competes with an interest of a person who consented to the change of law (subject, however, to the rule in Art. 7(5)).

7-19 Article 7(4)(a) provides that the existence of an interest in securities held with an intermediary that arose before the change of law is unaffected by the change, as well as the perfection (see commentary to Art. 1(1)(i) in paras. 1-20 and 1-21) of a disposition made prior to the change. However, while the perfected status of the interest acquired before the change of law continues after the change of law, it may be necessary by reason of Articles 7(4)(c) and 7(5) for the holder of that interest to take some action in order to preserve priority against a competing interest that arose before the change of law.

7-20 Article 7(4)(b) provides a rule similar to that in sub-paragraph (a) in specific situations.

7-21 Since some legal systems regard an attachment of securities or the opening of insolvency proceedings as equivalent to a disposition or as a process with similar effects, paragraphs 4(b)(ii) and (iii) prevent such events from being given the effect of a post-change disposition for purposes of Article 7 with the consequent possible effect of triggering the applicability of the new law. Thus, the old law continues to determine the legal nature and effects, including priority, of a pre-change interest in securities of a 'protected person' even against a person who attaches the securities *after* the change of law. Likewise, the old law continues to determine, with respect to a pre-change interest of a protected person, all the Article 2(1) issues with reference to an insolvency administrator in an insolvency proceeding opened *after* the change of law. This means that it is not open to the insolvency administrator to attack under the new law the existence or the perfection of an interest created or perfected before the change of law. These provisions ensure that the relations between a holder of an interest perfected before the change of law and creditors who attach after the change as well as creditors in insolvency proceedings that are commenced after the change, will be governed by the old law in spite of the change of law. Unrelated to the change of law, of course, the interest perfected before the change of law may be subordinated or set aside because of rules of insolvency relating to the ranking of claims or rules relating to avoidance (see Art. 8).

7-22 Article 7(4)(c) deals with the applicability of the old law to priority issues as between interests that arose before the change of law. Priority issues as between the interests in securities of protected persons which arose before the change of law are governed by the old law, subject to Article 7(5). In Article 7(4)(c) the term "arose" is to be understood as "perfected" because a priority issue presupposes the effectiveness of the security interests against third parties.

7-23 The new law generally does not affect the priority status *inter se* of persons who did not consent to the change of law. Where there are several pre-change security interests and some of the secured parties consented to the change of law while others did not, the priority rules of the new law apply only to disputes among the secured parties who did consent; the priority rules of the old law will apply to disputes between one or more parties who consented, on the one hand, and one or more parties who did not consent, on the other.

Example 7-5

An intermediary organised under the laws of Germany and its customer enter into an account agreement expressly providing that all the Article 2(1) issues are governed by Ruritanian law. The intermediary had a Qualifying Office in Ruritania at the time the governing law agreement was made. The account is credited with various securities. Subsequently the account holder pledges all the securities on Day 1 to a French bank and later, on Day 2, to an English bank and, on Day 3, to an Italian bank. All banks perfect, in the same order, the security interest in accordance with Ruritanian law by a method that does not result in a credit of the securities to the securities account of the collateral taking banks. Later, the German intermediary and its customer amend the account agreement so that it expressly provides that all the Article 2(1) issues are governed by Utopian law, and the intermediary had a Qualifying Office in Utopia at the time of the governing law amendment was made. French and Italian bank consent to the amendment, English bank does not. Later, in a proceeding in a State where the Convention is in force, a priority dispute arises among the three banks. The priority between English bank, on the one hand, and French and Italian bank, on the other, is determined under the old (Ruritanian) law. The priority between French bank and Italian bank is determined under the new (Utopian) law.

7-24 However, if some *dispositions* of the same securities occur before the change of law and others occur after the change of law, by virtue of Article 7(3) the new law governs the relative priority of *all* the interests (except as otherwise provided in Arts. 7(4)(c) and (5)). This assumes, of course, that none of the dispositions involves a transfer of securities from one securities account to another (see para. 7-1).

Example 7-6

An intermediary and its customer enter into an account agreement. The account agreement expressly provides that all the Article 2(1) issues are governed by Ruritanian law. The intermediary had a Qualifying Office in Ruritania at the time the governing law agreement was made. The account holder grants a security interest in the securities account and all the securities credited to the account. The secured party perfects the security interest in accordance with Ruritanian law by a method that does not result in a credit of the securities to its securities account. Later, the intermediary and its customer amend the account agreement so that it expressly provides that all the Article 2(1) issues are governed by Utopian law. The intermediary had a Qualifying Office in Utopia at the time of the amendment. The secured party did not consent to the amendment. Still later, an unsecured creditor of the account holder obtains a judicial order attaching all the securities credited to the securities account. Still later, in a proceeding in a State where the Convention is in force, the issues are (i) whether the security interest exists and is perfected; and (ii) the legal nature and effects (including the priority) of the security interest against the attaching creditor. The old law (Ruritanian law) determines all of these issues because the secured party did not consent to the change of law and, with respect to issues (i) because of Article 7(4)(a) and, with respect to issue (ii), because of Article 7(4)(b)(ii).

Example 7-7

An intermediary and its customer enter into an account agreement. The account agreement expressly provides that all the Article 2(1) issues are governed by Ruritanian law. The intermediary had a Qualifying Office in Ruritania at the time the governing law agreement was made. The account holder grants a security interest in the securities account and all the securities credited to the account. The secured party perfects the security interest in accordance with Ruritanian law by a method that does not result in a credit of the securities to its securities account. Later, the intermediary and its customer amend the account agreement so that it expressly provides that all the Article 2(1) issues are governed by Utopian law.

The intermediary had a Qualifying Office in Utopia at the time of the amendment. The secured party did not consent to the amendment. Still later, the account holder becomes insolvent. Still later, in an insolvency proceeding in a State where the Convention is in force, the question is which law governs all the Article 2(1) issues with respect to the interest of the secured party vis-à-vis that of the insolvency administrator. Subject to Article 8, the old law (Ruritanian law) governs all of these issues (including the priority of the interest of the secured party vis-à-vis that of the insolvency administrator), because the secured party did not consent to the change of law and because of Article 7(4)(b)(iii).

Example 7-8

An intermediary and its customer enter into an account agreement. The account agreement expressly provides that all the Article 2(1) issues are governed by Ruritanian law. The intermediary had a Qualifying Office in Ruritania at the time the account agreement was entered into. The account holder grants to secured party A (SP-A) a security interest in the securities account and all the securities credited to the account. SP-A perfects the security interest in accordance with Ruritanian law by a method that does not result in a credit of the securities to its securities account. Later, the account holder grants a security interest in the securities account and all the securities credited to the account to SP-B. SP-B perfects the security interest in accordance with Ruritanian law by a method that does not result in a credit of the securities to its securities account. Still later, the intermediary and its customer amend the account agreement so that it expressly provides that all the Article 2(1) issues are governed by Utopian law. The intermediary had a Qualifying Office in Utopia at the time of the amendment. Neither SP-A nor SP-B consented to the amendment. Still later, the account holder grants to secured party C (SP-C) a security interest in the securities account and all the securities credited to the account. SP-C perfects the security interest in accordance with Utopian law by a method that does not result in a credit of the securities to its securities account. Still later, in a proceeding in a State where the Convention is in force, the issues are (i) whether SP-A's security interest has priority over SP-B's security interest; and (ii) whether SP-A's and SP-B's security interests have priority over SP-C's security interest. The old law (Ruritanian law) determines issue (i) because both of the security interests arose before the change of law, neither secured party consented to the amendment and because of Article 7(4)(c); the new law (Utopian law) determines issue (ii) because the interest of SP-C arose after the change of law and, pursuant to Article 7(3), the new law governs all the Article 2(1) issues (and none of the exceptions in Art. 7(4) applies).

VI. Article 7(5) – Priority issues

7-25 Article 7(5) provides that, notwithstanding Article 7(4)(c), the new law governs the priority of an interest that arose under the old law, was not perfected under the old law but subsequently is perfected under the new law. Thus, by introducing the exception (Art. 7(5)) to the exception (Art. 7(4)), the general rule (*i.e.*, application of the new law, Art. 7(3)) applies. Article 7(5) applies solely in the case of security interests that were created under the old law merely between the parties, but subsequently are perfected under the new law. The question whether the old or the new law should govern the priority issue only arises in such "overlapping" cases (see Example 7-9). Where a security interest is perfected under the old law a repetition of such a perfection under the new law (may it occur before or after the change of law) does not trigger the applicability of the new law (see Example 7-10). As in such cases all security interests were perfected before the change of law the ranking of these security interests is governed by the old law. This reading is in conformity with the wording of Article 7(5) ("an interest that arose under the old law *but* is perfected under the new law"), which, in relation to a security interest, clearly opposes "arose" (under the old law) and "is perfected" (under the new law). This reading also prevents the legal security provided by Article 7(4)(c)) from being undermined by a collateral taker who triggers, by a repetitious perfection, the applicability of the new law in its favour.

Example 7-9

An intermediary organised under the laws of Luxembourg and its customer, a corporation incorporated under the laws of the State of New York, enter into an account agreement. The account agreement expressly provides that all the Article 2(1) issues are governed by Ruritanian law. The intermediary had a Qualifying Office in Ruritania at the time the governing law agreement was made. The account holder grants to secured party A (SP-A) a security interest in the securities account and all the securities credited to the account. SP-A perfects the security interest in accordance with Ruritanian law by a method that does not result in a credit of the securities to its securities account. Later, the account holder grants to SP-B a security interest in the securities account and all the securities credited to the account. SP-B, unlike SP-A, does not perfect the security interest in accordance with Ruritanian law. Still later, the intermediary and its customer amend the account agreement so that it expressly provides that all the Article 2(1) issues are governed by Utopian law. The intermediary had a Qualifying Office in Utopia at the time of the amendment. Neither SP-A nor SP-B consented to the amendment. Later, SP-B perfects the security interest in accordance with Utopian law. Still later, in a proceeding in a State where the Convention is in force, the issue is whether the new law (Utopian law) or the old law (Ruritanian law) governs whether the security interest of SP-B has priority over the security interest of SP-A. The new law (Utopian law) governs the priority issue because Article 7(5) is applicable since SP-B's security interest (which arose between the parties but was not perfected under the old law) is perfected only under the new law.

Example 7-10

An intermediary organised under the laws of Luxembourg and its customer, a corporation incorporated under the laws of the State of New York enter into an account agreement. The account agreement expressly provides that all the Article 2(1) issues are governed by Ruritanian law. The intermediary had a Qualifying Office in Ruritania at the time the governing law agreement was made.

The account holder grants to secured party A (SP-A) a security interest in the securities account and all the securities credited to the account. SP-A perfects the security interest in accordance with Ruritanian law by a method that does not result in a credit of the securities to its securities account. Later, the account holder grants to SP-B a security interest in the securities account and all the securities credited to the account. SP-B perfects in the same way as SP-A; however, SP-B also makes a filing at the filing office of Utopia. Still later, the intermediary and its customer amend the account agreement so that it expressly provides that all the Article 2(1) issues are governed by Utopian law. The intermediary had a Qualifying Office in Utopia at the time of the amendment. Neither SP-A nor SP-B consented to the amendment. Still later, in a proceeding in a State where the Convention is in force, the issue is whether the new law (Utopian law) or the old law (Ruritanian law) governs whether the security interest of SP-B has priority over the security interest of SP-A. As both of the security interests not only arose as between the parties but were perfected before the triggering amendment, the old law (Ruritanian law) governs the priority issue because Article 7(4)(c) is applicable since there is no "unperfected" situation as to a security interest. Independently of whether Utopian substantive law recognises an anticipated perfection (*i.e.*, a perfection made before Utopian law became applicable) by virtue of the filing, a repetitious perfection under the new law of a previously (under the old law) perfected security interest which is recognised by the new law cannot trigger the applicability of Article 7(5). The same is true when the repetition of the perfection takes place after the change of law by virtue of the triggering amendment. Otherwise the purpose of Article 7(4)(c) which is to provide legal certainty could easily be undermined.

Article 8 ***Insolvency***

1. **Notwithstanding the opening of an insolvency proceeding, the law applicable under this Convention governs all the issues specified in Article 2(1) with respect to any event that has occurred before the opening of that insolvency proceeding.**
2. **Nothing in this Convention affects the application of any substantive or procedural insolvency rules, including any rules relating to –**
   a) **the ranking of categories of claim or the avoidance of a disposition as a preference or a transfer in fraud of creditors; or**
   b) **the enforcement of rights after the opening of an insolvency proceeding**

I. Introduction

8-1 Article 8 addresses the question of the consequences of the opening of an insolvency proceeding for "any event" relating to securities held with an intermediary which occurred before the opening of that insolvency proceeding. Apart from an outright sale, the most important of these events is the grant of a security interest (*e.g.,* a pledge) whose effects against the intermediary and third parties are governed by the law applicable under the Convention (Convention law). The purpose of Article 8 is to ensure that pre-insolvency rights which are validly created and perfected under the Convention law are respected as such in an insolvency proceeding (Art. 8(1)), but that these rights are not thereby exempted from general insolvency rules (Art. 8(2); *e.g.*, as regards the ranking of claims, enforcement of rights, the avoidance of unfair preferences or transactions in fraud of creditors). Thus, the fundamental purpose of the Convention, that is, providing *ex ante* certainty with respect to the law governing all the Article 2(1) issues, in particular the creation and perfection of security interests, is upheld in the context of an insolvency proceeding; at the same time, the *lex concursus* is preserved. Thus, Article 8 fixes the limits between the Convention law (*lex causae*) and the applicable insolvency law (*lex concursus*). The Convention does not contain any rule on determining the *lex concursus*. It only prescribes in Article 8(1), that the latter, however determined, has to recognise legal situations which have occurred under the *lex causae* before the opening of the insolvency proceeding.

8-2 While the general approach of the Convention (recognition of interests acquired by virtue of the law applicable under the Convention even in an insolvency proceeding; in all other respects no interference with insolvency laws) was undisputed from the very beginning of the proceedings, the version of the provision finally adopted was fixed only after focused discussions (see Work. Docs. No 1, 5, and 9; Report of the Meeting No 7, paras. 1-43) in the course of the Diplomatic Conference. These discussions covered the need to clarify the scope and content of each paragraph as well as the relationship between them, in particular the boundary between the Convention law and the *lex concursus*.

8-3 When examining the potential impact of the insolvency proceeding on a security interest previously acquired by a secured creditor from the now insolvent debtor, two issues must be distinguished: the *recognition* of the rights acquired by the secured creditor (*e.g.*, respect for the perfected status of a pledge) (Art. 8(1); see paras. 8-7 and 8-8), and the *effects* of those rights in the insolvency proceeding of the debtor (*e.g.*, the vulnerability of those rights to the forum's insolvency rules and procedures generally applicable to all rights having such status) (Art. 8(2); see *paras.* 8-9 *et seq.*). The Convention distinguishes and addresses both these issues in Article 8. Article 8(1) establishes that the insolvency court has to unconditionally recognise a pre-insolvency interest perfected under the law applicable under the Convention. Article 8(2) makes clear that the consequences of that recognition are governed in any subsequent insolvency proceeding by the *lex concursus*.

II. Scope of Article 8 as to insolvency proceedings

8-4 Article 8(1) is drafted without specifying which insolvency proceedings are covered, that is, against whom the insolvency proceedings have to be brought in order that it is covered by the provision. Therefore, the provision applies with respect to insolvency proceedings against any party whose insolvency is relevant to the matter in question, whether the insolvent party is an account holder, a pledgee or transferee from an account holder, an intermediary or the issuer itself. The effect of Article 8(1) is that the perfected status of a pledge or transfer under the Convention law and the priority given to it by that law must be respected in the insolvency proceedings even if under the general (*i.e.*, non-insolvency) law of the jurisdiction in which those proceedings were opened the pledge or transfer would not have been treated as perfected or would have been accorded a different priority (see Examples 8-1 and 8-2). However, the rules of the relevant insolvency law may invalidate the pledge or transfer (*e.g.*, as an unfair preference or a transaction in fraud of creditors), may stay its enforcement, or may give it a lower ranking under those rules than it previously enjoyed (see Example 8-3 and Variant).

8-5 The only aspects of the situation when securities credited to a securities account are pledged by a pledgor who is not only an account holder but also an intermediary that differ from the standard situations discussed in paragraphs 8-1 to 8-3, 8-7 to 8-8, and 8-9 *et seq.*, are that, in the insolvency proceeding, (a) there may be a new class of claimants competing with the pledgee – the account holders of the now-insolvent intermediary, and (b) this new class, each member of which may base its claim on a different law because of the Convention rules, competes *inter se* as well as against the pledgee and other claimants. As discussed above, Article 8(2) preserves the freedom to apply the *lex concursus* with respect to the ranking of the perfected pledge vis-à-vis all competing claimants, including the pledgor's customers. While the Convention provides the conflict of laws rule with respect to the law governing the rights of each account holder against its intermediary, Article 8(2) preserves the applicability of the *lex concursus* to determine the relative rights among the account holders *inter se*.

8-6 Thus, the law designated by Article 4 or 5 will continue to determine whether an account holder has an effective right against its insolvent intermediary and whether a disposition in favour of a transferee from the intermediary has been perfected. However, in the intermediary's insolvency proceeding, Article 8(1) does not determine whether, when or how the account holder or the transferee with a perfected interest is able to realise that interest; nor does it determine the priority of distribution among account holders or the ranking of the positions of account-holders and transferees with a perfected interest. Article 8(2) expressly preserves the applicability of insolvency rules with respect to such issues.

Example 8-1

An investor organised under the laws of France and an investor organised under the laws Spain each hold, through an intermediary Ruritanian Bank, an interest in respect of 100,000 shares issued by a Company incorporated under the laws of Japan. Each of the two account agreements (effective to determine the applicable law under the Convention) designates as the governing law the law of the domicile of that investor. (This scenario is rather unlikely; it is far more likely that the intermediary will agree to only a single governing law in all of its customer account agreements; the example demonstrates the operation of the Convention even in an unusual case.) Ruritanian Bank holds those 200,000 shares through its own intermediary, Swiss Bank. Ruritanian Bank also owns outright 100,000 shares of the Japanese Company which it holds in its account at Swiss Bank. Ruritanian Bank pledges to Italian Bank the 300,000 shares reflected in its account at Swiss Bank. The pledge is perfected in accordance with Italian law, the law that governs the pledge under the rules of the Convention, based on the account agreement between Ruritanian Bank and Swiss Bank. An insolvency proceeding is opened in Ruritania against Ruritanian Bank.

Article 8(1) ensures that (1) the rights of each of the two investors as to its respective 100,000 shares vis-à-vis Ruritanian Bank continue to be governed by the law applicable under the Convention to each of their respective accounts, and (2) the perfected status of the pledge achieved under Italian law before the opening of the insolvency proceeding, with respect to all 300,000 shares, will be recognised as such by the Ruritanian insolvency court. Article 8(2) preserves the applicability of the Ruritanian *lex concursus*, so that it may produce the following three effects. First, the Ruritanian automatic stay may affect the pledge (despite its being perfected under Italian law) and the investors' accounts (despite being perfected under other non-Ruritanian laws). Secondly, Ruritanian insolvency ranking law may determine the ranking of the perfected pledge and the investors' accounts vis-à-vis competing claims. The competing claims might include (a) tax claims, (b) claims of other transferees from Ruritanian Bank, and (c) claims of Ruritanian Bank's customers (including the two investors). Thirdly, Ruritanian insolvency law may determine whether there are any applicable grounds for avoidance of the pledge. Under Article 8(2), the Ruritanian insolvency court is free to apply the *lex concursus* to determine the relative claims in the insolvency proceeding of the two investors *inter se* (suppose, for example, that there are only 100,000 shares available for distribution among the two investors), and this is true even if the competing investors have claims to their securities which are based on different State's laws. The foregoing example assumes that only one proceeding is opened. The Convention does not deal with the question which insolvency law (*e.g.*, with respect to the ranking of categories) will be applied by a particular insolvency forum when there are multiple or ancillary proceedings opened in several jurisdictions.

III. Article 8(1): Recognition of interests acquired prior to an insolvency proceeding

8-7 Article 8(1) responds to the concern that an insolvency administrator might apply to pre-insolvency interests the internal substantive law either of the forum State or of the State designated by the forum's conflict rules rather than the law applicable under the Convention. Hence, Article 8(1) provides that an interest acquired, prior to the opening of an insolvency proceeding,[31] under the Convention law must be recognised in the insolvency proceeding. This is perfectly sensible, as property rights cannot properly fulfil their function if the question of their creation and perfection can be differently answered depending on the State in which the insolvency proceeding is opened. While Article 8(1) covers all the Article 2(1) issues, it is mainly a rule of recognition whose effect is to prohibit an insolvency court from imposing any perfection rule other than that imposed under the Convention law. In other words, the insolvency court may not refuse to recognise the right or its perfected status merely because this right had not (also) been created or perfected in accordance with the (conflict or substantive) rules of the *forum concursus*. To this extent the Convention law prevails over the *lex concursus*. The use of the perfect past tense in the second half of Article 8(1) ensures that the provision applies only if the "event" (in practice, the most important of these "events" will be the crediting of securities to a securities account or the perfection of a disposition) has occurred before the opening of the insolvency proceeding. If the event occurs *after* the opening of an insolvency proceeding, Article 8(1) is not applicable. The preservation by Article 8(1) of the continued applicability after the opening of an insolvency proceeding of the Convention law is effective with respect to *all* the Article 2(1) issues.

8-8 This rule of recognition does not operate in a vacuum, but rather must be understood in the context of the sphere of insolvency law as preserved in Article 8(2), that is, the effects of such recognised rights in the insolvency proceeding are governed by the *lex concursus*. Thus, the Article 8(1) rule of recognition does not entail that the rights so recognised are immune from the application of *lex concursus* rules generally applicable to such rights in insolvency proceedings. This contextual general rule, set forth in Article 8(2), is discussed more fully below (see paras. 8-9 *et seq.*).

Example 8-2

> An investor has securities credited to a securities account maintained for it by an intermediary; the account agreement expressly provides that it is governed by the laws of Luxembourg and does not expressly provide that a different law governs all the Article 2(1) issues. The intermediary had, at the time the governing law agreement was made, a Qualifying Office in Luxembourg. The securities are pledged to Bank organised under the laws of Italy. The pledge is, under the

---

31 The moment of the *opening* of an insolvency proceeding is determined in accordance with the *lex concursus*. The state of insolvency will usually (in some countries, must) precede the opening of an insolvency proceeding; it is the latter event that is the relevant time for Art. 8 (*e.g.*, time when opening is registered with a public record or time at which a judgement opening proceedings becomes effective).

Convention, governed by, and it was perfected in accordance with, Luxembourg law. An insolvency proceeding is opened in London against Investor. Article 8(1) ensures that the pledge perfected under Luxembourg law before the opening of the insolvency proceeding will be recognised as such by the English insolvency court.

IV. Article 8(2): Effects of previously acquired interests in an insolvency proceeding

8-9 The preservation of the proper role of the Convention law is achieved by Article 8(1). Article 8(2) preserves the proper role of the *lex concursus* with respect to the previously acquired rights recognised pursuant to Article 8(1). Article 8(2) ensures that although the law applicable under the Convention governs the existence of the rights, it is left to the *lex concursus* to determine the effects of those rights (*i.e.*, the extent to which the secured party can actually use its rights in the insolvency proceeding).

8-10 To define clearly and to underscore the applicability of the *lex concursus*, Article 8(2) is structured to make the same point in two different ways. First, the chapeau lays down the principle – that "[n]othing in this Convention affects the application of any substantive or procedural insolvency rules" – in sweeping unqualified language. (Of course, in spite of such general wording, paragraph 2 has to be read together with paragraph 1 and does not impair the latter; for convenience of the reader, paragraph 2 could have started: "Without prejudice to paragraph 1, nothing … ".) Secondly, using the term "including" to negate the implication of a limitation of the scope, the text gives a number of examples of insolvency rules that are not affected by the Convention: "[…] including any rules relating to […]". Thus, Article 8(2)(a) expressly preserves rules of the *lex concursus* relating to the *avoidability* of fraudulent transfers and preferences (*e.g.*, where, on the eve of an insolvency proceeding, the debtor grants an interest to a creditor which gives the latter an impermissible preference at the expense of other creditors), as well as rules which may declare that certain types of claims (*e.g.*, for wages and/or taxes) should have priority over any other interest. Furthermore, to ensure that reorganisation or bankruptcy proceedings do not collapse (because, *e.g.*, a secured party seizes crucial assets), Article 8(2)(b) subjects the remedies otherwise available to even a recognised perfected security interest to rules generally applicable in insolvency proceedings designed to prevent such disruption (*e.g.*, a stay). The nature and procedure of such mechanisms are subject to the *lex concursus*. For example, it is open to the *lex concursus* to exempt certain securities (*e.g.*, those that are listed on a stock exchange) from the scope of a general stay provision. In sum, the opening of an insolvency proceeding against an account holder does not change the law determined under Article 4 or 5 to govern the nature and perfected status of the pledgee's or transferee's interest, but the appropriate insolvency law will determine, for example, whether that pledge is, nevertheless, avoidable as a preference or whether its enforcement is subject to a stay.

8-11 The statement in Article 8(2) that nothing in the Convention "*affects*" the application of avoidance and other insolvency rules is to be understood as a statement that the Convention *does not deal with the law applicable to those issues* and, as a consequence, has no conflict of laws impact on those issues.

8-12 The chapeau of Article 8(2) expressly specifies that not only *procedural* insolvency rules are preserved, but also *substantive* insolvency rules. The relevant rules are characterised as procedural rules in some systems and as substantive rules in others. The Convention text ensures that the reservation in favour of the insolvency law is not artificially restricted by differing understandings of the terms 'procedural' and 'substantive'. This seems to be particularly important with respect to avoidance rules, which might be considered procedural rules in some systems and substantive rules in others.

Example 8-3

C made an unsecured loan to D of $1 million. Nine months later, C becomes concerned that D might soon become insolvent and takes a security interest of D's rights in securities credited to a securities account to secure the loan. The security interest is, under the Convention, governed by the law of Utopia, and it is created and perfected in accordance with that law. One month later, a court in Ruritania issues a winding-up order against D on the ground of insolvency and appoints an insolvency administrator. Under Ruritanian insolvency law, a security interest given to secure antecedent indebtedness (past value under Ruritanian law) within the six-month period preceding the opening of an insolvency proceeding may be set aside as a preference on the application of the insolvency administrator. If D's insolvency administrator applies to set aside the security interest given to C, even though the security interest was perfected in accordance with the law of Utopia prior to the opening of the insolvency proceeding and even though that perfected security interest interest is recognised by the insolvency court (as required under Art. 8(1)), that recognised perfected status will not provide a defence to the application to set aside the security interest under Ruritanian insolvency rules concerning the avoidance of preferential transfers; such recognition (as preserved under Art. 8(1)) does not preclude application of the avoidance rules of the *lex concursus* (as preserved under Art. 8(2)).

Variant:

C made a loan to D and took a security interest of D's rights in securities credited to a securities account to secure the loan. The security interest is, under the Convention, governed by the law of Utopia, and it is created and perfected in accordance with that law. Seven months later, a court in Ruritania issues a winding-up order against D on the ground of insolvency. Under Ruritanian insolvency law (in contrast to Utopian insolvency law) any security interest ranks behind a legal governmental lien for taxes. Thus, even though C's security interest was perfected in accordance with the law of Utopia seven months prior to the opening of the insolvency proceeding and even though that perfected security interest is recognised by the insolvency court (as required under Art. 8(1)), that recognised perfected security interest will be given a lower rank in the insolvency than it would have had under the (general) Convention law (Art. 8(2)).

CHAPTER III GENERAL PROVISIONS

Article 9 *General applicability of the Convention*

This Convention applies whether or not the applicable law is that of a Contracting State.

9-1 Article 9 follows the pattern of earlier Hague Conventions and of the *Rome Convention of 1980 on the Law Applicable to Contractual Obligations* in making clear the universal character of the conflict rules in the Convention. The law determined by the Convention is the applicable law whether or not it is the law of a Contracting State (on the meaning of the term "Contracting State", see para. 1-43), whether or not it is the law of a Member State of the Hague Conference on Private International Law, and, if the law is that of a territorial unit, whether or not the territorial unit is one to which the Convention has been extended under Article 20. Thus, the effect of Article 9 is to preclude an inappropriate limitation of the reach of the conflict of laws regime provided in Articles 4 and 5. Article 9 applies equally if the law determined under the Convention is the law of a Regional Economic Integration Organisation to whom competence has been given by its sovereign member States and whose law is that equated with that of a Contracting State under Article 18(3).

Article 10 ***Exclusion of choice of law rules (renvoi)***

In this Convention, the term "law" means the law in force in a State other than its choice of law rules.

10-1 Article 10 makes clear that the conflict of laws rules of the Convention refer to domestic (substantive) rules only, not to conflict of laws rules. The Convention does not leave any room for *renvoi* in the traditional private international law sense. Thus, the Convention ensures that there will be no further reference by the Convention law (whether that is the law of a State or of a territorial unit) to the law of another State or the law of a territorial unit of another State (whether the law designated stems from the State, the territorial unit or a Regional Economic Integration Organisation). Article 10 is a core provision of the Convention since it guarantees the achievement of the very purpose of the Convention (*ratio conventionis*), that is, the unification of diverging national conflict of laws rules in the interest of certainty, predictability and simplicity. Since, under Article 9, the Convention applies whether or not the applicable law is that of a Contracting State, the admission of *renvoi* would frustrate the unifying purpose of the Convention if its conflict of laws rules point to the law of a State where the Convention is not in force and whose conflict of laws rules do not coincide with those of the Convention. Accordingly, the provision on the exclusion of *renvoi*, adopted by the Diplomatic Conference, has been uncontroversial from the very beginning of the proceedings (see Prel. Doc. No 1, p. 41 and Art. 8 of the January 2001 draft) and is in line with applicable law provisions in modern Hague Conventions, which, in general, exclude *renvoi*.

10-2 However, in the intention not to interfere with internal law, the Convention provides in two instances (Art. 12(2)(b) and (3)) for a form of internal *renvoi* within a Multi-unit State (see commentary to Art. 12).

10-3 If parties to an account agreement fail to select a substantive governing law directly and instead specify that their account agreement or all the Article 2(1) issues are governed by the law determined by the conflict of law rules of a particular jurisdiction, Article 10 prevents such a clause from being a governing law agreement under Article 4. Consequently, the relevant fall-back rule in Article 5 applies. Otherwise the parties would be allowed to undermine the unifying purpose of the Convention.

10-4 The phrase "law in force" is used in preference to "law of" to cover cases where, in a territorial unit, the relevant law includes both the law of that territorial unit and, to the extent applicable (under the law of either the territorial unit or the Multi-Unit-State) in that unit, the law of the Multi-unit State (see also paras. 4-15, 12-12 and 12-13).

Article 11 ***Public policy and internationally mandatory rules***

1. **The application of the law determined under this Convention may be refused only if the effects of its application would be manifestly contrary to the public policy of the forum.**
2. **This Convention does not prevent the application of those provisions of the law of the forum which, irrespective of rules of conflict of laws, must be applied even to international situations.**
3. **This Article does not permit the application of provisions of the law of the forum imposing requirements with respect to perfection or relating to priorities between competing interests, unless the law of the forum is the applicable law under this Convention.**

I. Introduction

11-1 Article 11 defines and carefully restricts the grounds for judicial refusal to apply the law determined under either Article 4 or 5 of the Convention. By doing so, Article 11 not only supports the Convention's goal of providing legal certainty and predictability but also establishes the importance that Contracting States attribute to this goal. In sum, under Article 11, courts should only in extraordinarily rare cases avoid the application of the law determined under the Convention (the Convention law).

11-2 While the first draft (January 2001 draft; Prel. Doc. No 2) contained two provisions on the issue (*i.e.*, Art. 9 on internationally mandatory law of the forum, which already did not permit the application of non-Convention law provisions relating to perfection and priorities, and Art. 10 on public policy), in the following drafts these two Articles were combined in one provision (see Art. 8 of the November 2001 draft; Prel. Doc. No 6). Thereafter, only minor drafting changes occurred.

11-3 The three paragraphs in Article 11 set forth the framework for safeguarding other goals of the public policy of a Contracting State in the context of the goal of legal certainty. Article 11(1) may be regarded as addressing the "defensive" or "negative" facet of public policy; it provides a mechanism which might, under strict conditions, lead to the *refusal* by a court to apply specific rules of the Convention law because the effects of its application would be manifestly contrary to the public policy of the forum State. In contrast, Article 11(2), which provides for the application of the forum's internationally mandatory rules, may be regarded as addressing the "offensive" or "positive" facet of public policy, that is, a mechanism which *requires* that a specific norm of the forum State be from the outset applied despite the fact that the Convention determines another law to be applicable and irrespective of the effects the latter would have. Although these two mechanisms differ in their approaches, they pursue the same purpose: the safeguard of fundamental moral, social, economic or political principles of the *forum* State. In light of their close correlation and because both are subject to the limitation in Article 11(3), the two exceptions have not been separately stated in different articles as in some other Hague Conventions (see, *e.g.,* Arts. 17 and 18 of the *Hague Convention of 1986 on the Law Applicable to Contracts for the International Sales of Goods*). Both exceptions are substantially qualified by Article 11(3) –

arguably the most important provision in Article 11. Article 11(3) effectively puts an important limit on the two exceptions by stating that they may not be used to impose requirements with respect to perfection or relating to priorities between competing interests, unless the law of the forum is the Convention law (see full discussion in para. 11-12).

11-4 Article 11 should not be brought into play in the *insolvency* context, as the Convention, in Article 8, provides specific rules with respect to insolvency law (*lex concursus*), thus precluding resort to the general public policy provision of Article 11 as the basis for application of the forum's insolvency law. Article 11 only concerns the relationship between the (substantive) law of the forum (*lex fori*) and the (substantive) law applicable under the Convention (*lex causae*). The relationship between *lex causae* and *lex concursus* is outside the scope of Article 11 but covered by Article 8. Article 11 therefore neither expands nor contracts the scope of the *lex concursus* as defined in Article 8. Hence, it is clear that Article 11 does not prohibit the application of *insolvency* rules of the forum, permitted under Article 8, establishing priorities between competing interests relating to the same securities. If the forum State is also the State of opening of an insolvency proceeding, the application of either Article 8 or Article 11 depends on whether the relevant provision forms part of the insolvency law or not.

11-5 While the application of national rules on "fraud of law" is not *in haec verba* prevented by the Convention, there seems, *de facto*, no room for the application of such provisions in the context of the Convention as the Convention's conflict of laws rules are preventing, by themselves, fraudulent determination of the applicable law (see paras. 3-10, 4-7 and 4-21 *et seq.*).

II. Article 11(1): Public policy exception

11-6 Article 11(1) sets forth a public policy *(ordre public)* exception to the application of the law otherwise applicable under the Convention; it is similar to public policy provisions found in most private international law treaties. As is the case for all modern Hague Conventions, the exception states that the law determined to be applicable under the Convention may be ignored only if its application is *manifestly* incompatible with the public policy of the *forum*. There is no definite formula describing the precise content, or the extent of the stringency, of the State's public policy which is necessary in order to permit (indeed, require) a court to refuse to apply the law otherwise applicable under the Convention. The exception must be applied on a case-by-case basis. This is not to suggest, however, that the matter is left to the discretion of the particular judge – to do so would undermine the basic goal of the Convention (*i.e.*, *ex ante* certainty) and would be contrary to the policy adopted by the Legislator (or whichever organ acts on behalf of a State or Regional Economic Integration Organisation) in deciding to become a Party to the Convention. Rather, there is abundant authority to the effect that the public policy exception applies only in the admittedly extremely rare cases where the relevant foreign rule, as applied to the facts of the case, would produce a result which departs so radically from the forum's concepts of fundamental justice that its application would be intolerably offensive to the forum's basic values.

11-7 Clearly, such a level of fundamental departure from basic values is *not* reached when, for example, the legal nature of the rights resulting from a credit of securities to a securities account is merely contractual under the Convention law, while under the law of the forum these rights would be proprietary in nature or vice versa. Likewise, such a level of fundamental departure would not be reached if either law classified the transfer of title by way of security or a sale and repurchase agreement as an effective disposition, whereas the other law would classify that transaction as an ineffective or unperfected disposition, or either law provided that the effects of the rights acquired by a person from a disposition of securities included the right to reuse the securities with or without the consent of the other person, whereas the other law would treat the legal nature and effects of a disposition as being more limited (see paras. 2-18 to 2-20). Moreover, as Article 11(3) expressly makes clear, foreign perfection requirements or priority rules do not violate the public policy of the forum merely because they are different from the forum's requirements or rules.

11-8 As stated in the introductory comments (see para. 11-3), although Article 11(1) is drawn up in standard Hague Convention terms, it must be read in light of the important Article 11(3) to which Article 11(1) is subject (see para. 11-12).

III. Article 11(2): Internationally mandatory rules of the forum

11-9 Article 11(2), as some provisions in other private international law treaties (*e.g.*, Art. 7(2) of the *Rome Convention of 1980 on the Law Applicable to Contractual Obligations*), provides that the Convention does not prevent the application of the forum's rules that are "mandatory" in the private international law sense. The rules referred to in Article 11(2) are *substantive* rules (*i.e.,* not private international law rules) that must be applied even to transactions that are transnational or wholly foreign, this compulsion being explicit in the law or implicit (based upon an underlying fundamental purpose the achievement of which requires application even to non-domestic transactions governed by foreign law). The rules referred to are not merely mandatory in the internal law sense of being imperative and not subject to variation by contract, but must be mandatory in the private international law sense that the forum court is bound to apply them even in non-domestic transactions governed by foreign law. In other words, internationally mandatory rules of the forum are such *substantive* provisions which are to be *exclusively* applied even when the forum's rules of private international law designate a foreign legal system as applicable and irrespective of the content of the latter, that is, even when the result of the application of the (overruled) pertinent rules of the designated foreign law would have been the same as under domestic law. It is clear that only the forum's internationally mandatory *rules* are to be applied, while, for the rest, the Convention law continues to govern. Again, this is not a matter left to the discretion of the particular judge, and, again, it is anticipated that application of this exception will be extremely rare.

11-10 The Article 11(2) exception must be read in light of Article 11(3) to which it is subject (see para. 11-12).

11-11 As Article 11(2) clearly states, this exception applies only to internationally mandatory rules *of the forum,* but not to those of other States. Thus, internationally mandatory rules of *third States* are *not* to be applied or to be taken into account. In this respect clear and swift consensus among the experts existed from the very beginning of the proceedings (see Prel. Doc. No 1, p. 48; Prel. Doc. No 2, p. 31). Expansion of this exception in favour of internationally mandatory rules of third States would potentially reintroduce uncertainty (*e.g.*, the rules of which third State should be applied? Which of the third State's rules are internationally mandatory? Should the application be compulsory or discretionary?) and thus frustrate the purpose of the Convention (*i.e.*, legal certainty). Moreover, the discussion of the expansion concept noted that, although the expanded exception is admitted in some countries in a contractual context, the concept of internationally mandatory rules of third States is not appropriate in the context of the Article 2(1) issues.

Example 11-1

> Investor I, a resident of State A, has securities credited to a securities account maintained for it by an intermediary which does a substantial volume of pertinent business in State D; the securities credited to the account are issued by a corporation incorporated under the laws of State C. Assume that under the Convention (whether its primary rule or one of its fall-back rules), the law governing all the Article 2(1) issues is that of State B. Investor I grants a security interest in the securities. Relying on the Convention, the secured party perfects its security interest in accordance with the law of State B. The law of State A requires that pledges of securities be perfected by a filing in a public register in State A. The law of State C requires that all pledges of securities issued by corporations incorporated in State C be noted on the books of the corporation or those of its registrar. The law of State D requires that all pledges of securities be perfected by means of a notarial instrument. The law of State E requires that all pledges of securities be perfected by publication of a notice posted on the courthouse door. Litigation concerning the perfection and priority of the pledge occurs in State E, although State E has no connection with the parties or the pledge. As the Convention is in force in State E, the law of State B governs all the Article 2(1) issues. In the litigation, a competing claimant asserts that the pledge is unperfected under the laws of States A, C, D and E. Under the Convention's exception in Article 11(2), the court can take into account only the rules relating to perfection of State B (see para. 11-12) and not those of States A, C, D or E, even if the relevant rules of those States were expressly internationally mandatory and even though each of States A, C and D has some connection with the securities or the parties.

IV. Article 11(3): An important limitation

11-12 Article 11(3) makes clear that even the public policy exception (see paras. 11-6 to 11-8) and the internationally mandatory forum rule exception (see paras. 11-9 to 11-11) may not be used to impose any perfection requirements of the forum (*e.g.*, the requirement under the UK Companies Act 1985 to register certain types of charges) on any consensual security interest or other disposition (incl. any statutory lien falling within Art. 1(2)(c)) in lieu of or in addition to the perfection requirements of the Convention law. Nor may these exceptions be used to apply any rule of the forum relating to priorities between competing interests. Article 11(3) thus ensures that the forum's law (i) imposing requirements for perfection or (ii) relating to priorities does not displace the Convention law with respect to the perfection of interests or priorities between competing interests, under the guise of applying the public policy exception or internationally mandatory rules. Article 11(3) is thus consistent with and reinforces Article 8(1), which prevents an insolvency court from imposing any perfection requirements other than those imposed under the Convention law with respect to any disposition that occurred before the opening of a particular insolvency proceeding (see paras. 8-7 and 11-4). Article 11(3) does not, however, make the public policy or internationally mandatory rules exceptions (Arts. 11(1) and (2)) inapplicable to a priority contest involving a governmental claim of the State which is the forum (*e.g.*, a tax lien) or to a situation in which the application of the Convention law would conflict with the forum's regulatory laws, such as anti-money laundering or anti-tax evasion laws, which do not create competing interests but instead regulate behaviour. By becoming a party to the Convention, a State indicates its determination that the widespread – virtually universal – utilisation of the Convention's conflict of laws rules, even when that results in the application of perfection and priority rules different from its own substantive rules, promotes economic stability and growth of that State (as well as globally) and also redounds to the benefit of parties (incl. governmental institutions and financial institutions vital to the State) within as well as outside that State. Article 11(3) reflects the decision by each Contracting State that the public policy of that State, with respect to perfection and priorities, is the application of the substantive law determined under the Convention, not the respective substantive rules of the forum, and that individual judges are not to frustrate attainment of the goal of *ex ante* certainty by invoking public policy considerations to give primacy to another policy so as to change that result. Article 11(3) is not at all a derogation from the Contracting State's public policy; rather, it is a clear announcement of what that policy is.

Example 11-2

> If a collateral taker has perfected, under Singapore law, its security interest in securities issued by an issuer organised under the laws of Malaysia, and Singapore law is the law determined under the Convention as the applicable law governing all the Article 2(1) issues, the security interest will be recognised as perfected in all fora of States where the Convention applies. In other words, any perfection requirement imposed by a law other than Singapore law need not be satisfied before a court of a State where the Convention applies. This is the case even if Malaysian law includes special regulatory requirements or procedures regulating the transfer of securities on the books of the Malaysian CSD (including, *e.g.*, provisions prohibiting the holding of securities in off-shore accounts). Article 11(3) ensures that such regulatory requirements cannot add to or displace the Singapore perfection law and must be ignored by the courts of all States party. In other words, while such regulatory schemes clearly remain the prerogative of any Contracting State, they cannot be regarded as a necessary further step for 'perfection'. This is particularly appropriate since the most likely fact pattern presenting this issue would involve transfers operated on the books of lower-tier intermediaries abroad who have agreed with their account holders that the governing law is a law other than Malaysian law, but the rule is by no means limited to that pattern. Of course, if the proceeding takes place in a court of Malaysia and if but only if Malaysian law is the applicable law under the Convention which governs the perfection with respect to this security interest in these securities, the Malaysian judge will apply Malaysian law and may regard the interest of the collateral taker as unperfected (or possibly void) if any applicable

Malaysian requirements have not been satisfied. In such a case, however, it will do so because Malaysian law is the Convention law, not because of the application of those requirements under Article 11(1) or (2) as a matter of Malaysian public policy. As a result of Article 11(3), even in a Malaysian court, the Malaysian regulatory requirements cannot be imposed as additional requirements for perfection.

Article 12 ***Determination of the applicable law for Multi-unit States***

1. If the account holder and the relevant intermediary have agreed on the law of a specified territorial unit of a Multi-unit State –

a) the references to "State" in the first sentence of Article 4(1) are to that territorial unit;

b) the references to "that State" in the second sentence of Article 4(1) are to the Multi-unit State itself.

2. In applying this Convention –

a) the law in force in a territorial unit of a Multi-unit State includes both the law of that unit and, to the extent applicable in that unit, the law of the Multi-unit State itself;

b) if the law in force in a territorial unit of a Multi-unit State designates the law of another territorial unit of that State to govern perfection by public filing, recording or registration, the law of that other territorial unit governs that issue.

3. A Multi-unit State may, at the time of signature, ratification, acceptance, approval or accession, make a declaration that if, under Article 5, the applicable law is that of the Multi-unit State or one of its territorial units, the internal choice of law rules in force in that Multi-unit State shall determine whether the substantive rules of law of that Multi-unit State or of a particular territorial unit of that Multi-unit State shall apply. A Multi-unit State that makes such a declaration shall communicate information concerning the content of those internal choice of law rules to the Permanent Bureau of the Hague Conference on Private International Law.

4. A Multi-unit State may, at any time, make a declaration that if, under Article 4, the applicable law is that of one of its territorial units, the law of that territorial unit applies only if the relevant intermediary has an office within that territorial unit which satisfies the condition specified in the second sentence of Article 4(1). Such a declaration shall have no effect on dispositions made before that declaration becomes effective.

I. Introduction

12-1 Article 12 sets forth several crucial interpretative and substantive provisions relating to the application of the Convention with regard to Multi-unit States. The Convention defines a Multi-unit State as "a State within which two or more territorial units of that State, or both the State and one or more of its territorial units, have their own rules of law in respect of any of the issues specified in Article 2(1)" (see commentary on Art. 1(1)(m) in para. 1-28). The sets of rules, therefore, may emanate either from different territorial units of a Multi-unit State (*e.g.*, the states and territories of Australia), or from different levels of government (*e.g.*, federal and state laws in the United States).

12-2 Article 12 is not meant to be applied exclusively by or within Multi-unit States, but is equally important for non-Multi-unit States; it must be given effect by and within *all* States where the Convention is in force, in determining, with regard to Multi-unit States, the law applicable to any of the issues specified in Article 2(1).

12-3 Article 12(1) explicates how the primary rule (Art. 4) operates in relation to Multi-unit States when the parties have designated the law of a particular territorial unit (see paras. 12-6 *et seq.*); Article 12(2)(a) clarifies the meaning of the phrase "law in force in a territorial unit of a Multi-unit State" used in the Convention (see paras. 12-12 and 12-13); Article 12(2)(b) preserves internal conflict of laws rules in respect of perfection by filing, recording or registration (see paras. 12-14 and 12-15); and Article 12(3) and (4) permit a Multi-unit State to make certain declarations in relation to the operation of Articles 4 and 5 in that Multi-unit State (see paras. 12-16 *et seq.*, and 12-21 *et seq.*).

12-4 Article 12 is *not* a Multi-unit State *extension* clause (see commentary to Art. 20).

12-5 The final version of Article 12 could only be drafted after the preliminary question of how the conflict choice of laws rules would look was definitely answered. This answer was given only during the Diplomatic Conference but in a way allowing considerable simplification of the very complex last pre-Conference draft (see Art. 11, June 2002 preliminary draft, Prel. Doc. No 15). With regard to this genesis of Article 12, most of the drafts and materials produced during the proceeding are not very helpful for the interpretation of that provision.

II. Article 12(1): The operation of the Convention's primary rule (Art. 4(1)) in relation to a Multi-unit State

12-6 Article 12(1) provides rules for interpreting Article 4(1) when, but only when, the account holder and the relevant intermediary have agreed on the law of a specified territorial unit of a Multi-unit State. In this important respect, paragraph 1 differs from the remainder of Article 12 whose provisions are not dependent on such an agreement. It is clear that the applicability of Article 12(1) presupposes the agreement under Article 4(1) being effective.

12-7 If the parties have not designated the law of a particular territorial unit, but instead have either (a) not expressly agreed on a law governing the account agreement or all of the issues specified in Article 2(1), or (b) expressly agreed on the law of a Multi-unit State itself (*e.g.*, the "law of Canada" or the "law of the United States") (this latter event is unlikely to occur if either of the parties is well-advised), paragraphs (1) and (4) are not applicable; the remainder of Article 12, however, nevertheless applies whenever the applicable law determined by the Convention is that of a Multi-unit State.

12-8 Article 4(1) only gives effect to the express agreement in an account agreement to the law of a *State* provided that the relevant intermediary has a Qualifying Office in that *State.* Thus, Article 4(1) presupposes that the State pointed to has only one set of pertinent rules. Article 4(1) does not address the situation in a Multi-unit State. Hence, Article 12(1) is designed to adapt Article 4(1) to such situations and, by doing so, the provision implicitly states that the parties may choose the law not only of *States* but also of *territorial units.* As Article 4(1) uses the term State several times, Article 12(1) explicates how to construe the term "State" in these several references and establishes that, with respect to Multi-unit States, the term "State" in Article 4(1) has two different meanings.

12-9 First, paragraph (1)(a) makes clear that if the parties have expressly agreed on the law of a particular territorial unit of a Multi-unit State (either as the law governing the account agreement or as another law governing all the Article 2(1) issues), the applicable law is the law of that specified territorial unit (subject to the Qualifying Office condition, as appropriately applied in this context; see para. 12-10). In other words, if the parties designate the law of a particular territorial unit, their agreement does not reach merely the Multi-unit State but indeed reaches the specified territorial unit. For example, if the parties have agreed on the law of New South Wales (NSW), paragraph (1)(a) provides that the law in force in NSW will apply to all the Article 2(1) issues; thus, in the context of the parties' specification of the law of a particular territorial unit of a Multi-unit State, the references to "State" in the first sentence of Article 4(1) are references to that territorial unit (see paras. 12-12 and 12-13, explaining what is actually meant by a reference to the "law in force" in a territorial unit, and paras. 12-14 and 12-15, on the preservation of internal conflict of laws rules in respect of perfection by filing, recording or registration).

12-10 On the other hand, in this same scenario (*i.e.*, if the parties have agreed on the applicability of the law of a specified territorial unit of a Multi-unit State), paragraph (1)(b) establishes that the term "that State" in the second sentence of Article 4(1) refers to the Multi-unit State itself. Thus, the Qualifying Office condition is satisfied if the relevant intermediary has a Qualifying Office *anywhere* in the Multi-unit State (see Example 12-2). Thus, the Qualifying Office need not necessarily be in the territorial unit whose law has been agreed upon by the parties. Paragraph (4), however, permits, by means of a declaration, a Multi-unit State to impose (to be enforced in every State where the Convention is in force, not just when that Multi-unit State is the forum) a geographically more stringent condition than the one generally applicable under Article 4(1) (see paras. 12-21 *et seq.*).

12-11 With respect to paragraph (1)(b), two further observations are to be made. First, the interpretative rule of paragraph (1)(b) *cannot* lead to a *renvoi* situation, that is, to the result that the applicable law is that of either (a) a State other than the State within which the agreed territorial unit is located, or (b) a territorial unit within a Multi-unit State other than the Multi-unit State within which the agreed territorial unit is located. This is true both because of the express language of paragraph (1)(b) and because of Article 10 which generally excludes *renvoi* in the Convention. Secondly, the interpretative rule of paragraph (1)(b) does *not* imply that parties that agree on the law of a particular territorial unit of a Multi-unit State do not have to satisfy the Qualifying Office requirement while parties that agree on the law of a non-Multi-unit State would have to satisfy that requirement. Neither set of parties is free from the need to satisfy the requirement that there is situated, at the time the governing law agreement is made, a Qualifying Office somewhere in the relevant State (Multi-unit State or not). It is the Multi-unit State itself that determines, by making or not a declaration under paragraph (4), whether, as a matter of its internal policy, the Qualifying Office need be within the agreed territorial unit or may be anywhere in the Multi-unit State; the parties must satisfy the Qualifying Office requirement in this way determined by the Multi-unit State.

Example 12-1

Investor I opens a securities account with an intermediary organised under the laws of the state of Ohio, Bank A. The account agreement expressly specifies that it is governed by the laws of the state of New York and does not expressly provide that a different law governs all the Article 2(1) issues. At the time the governing law agreement was made, Bank A had a Qualifying Office in New York. Later, Investor I grants to a lender organised under English law a security interest in Investor I's securities account and all securities credited to it. Under Article 4(1), first sentence, as interpreted using the rule of Article 12(1)(a), all the Article 2(1) issues, including the perfection issue, are governed by the law in force in the state of New York. In this example, a not uncommon scenario, the New York office would have satisfied the Qualifying Office requirement, whether or not the United States had made a declaration under Article 12(4), because the office was located in the state of New York, the territorial unit designated by the governing law clause. Because the state of New York is a territorial unit of a Multi-unit State, Article 12(2)(a) will be relevant in determining the content of that law, and, under some circumstances, Article 12(2)(b) will also be relevant (see paras. 12-12 and 12-13, and 12-14 and 12-15 respectively).

Example 12-2

Same facts as Example 12-1, except that at the time the governing law agreement was made, Bank A's only office was in the state of Ohio. If the United States has not made a declaration under Article 12(4), the condition specified in the second sentence of Article 4(1) is nevertheless satisfied, as Bank A had a Qualifying Office within the Multi-unit State, that is, within the United States (even though that office was not within the territorial unit specified in the account agreement). Thus, as in Example 12-1, all the issues specified in Article 2(1) are governed by the law in force in the state of New York. Again, paragraph (2)(a) will, and paragraph (2)(b) might also, be relevant.

Example 12-3

An intermediary and its customer enter into an account agreement which expressly provides that it is governed by the laws of British Columbia and does not expressly provide that a different law governs all the Article 2(1) issues. At the time the governing law agreement was made, the intermediary had no office in British Columbia or anywhere else in Canada. The condition specified in the second sentence of Article 4(1), even though interpreted pursuant to Article 12(1)(b), is not satisfied, as the intermediary has no Qualifying Office in Canada. The parties' designation of the law of British Columbia as the governing law is thus not effective under Article 4(1) to determine the law governing all the Article 2(1) issues. As a result, if later the shares credited to the customer's securities account are pledged, the applicable law is not determined by Article 4 but by one of the fall-back rules in Article 5. Note that this might bring Article 12(3) into play, if the fall-back rule designates the law of a Multi-unit State and that State has made a declaration under that paragraph.

III. Article 12(2)(a): The "law in force in a territorial unit of a Multi-unit State"

12-12 Paragraph (2)(a) clarifies the meaning of the term "law in force in a territorial unit of a Multi-unit State", a term used in several provisions of the Convention (Arts. 4(1), 5(1), 5(2), 5(3), 10, 12(2)(b), 16(3), and 16(4)). Paragraph (2)(a) defines this term as including both the law of that *territorial unit* and, to the extent applicable (under the law of either the unit or the Multi-Unit-State) in that unit, the law of the *Multi-unit State*. This rule recognises and automatically implements federalism rules that exist in some Multi-unit States (for example, preserving the effect of prevailing federal rules applicable to U.S. Government and Agency securities). Thus, if, under the relevant Convention rule, the applicable law is, for example, New York law, this means not only New York state law but also any US federal law in force in New York which is relevant to the issue in question.

12-13 Paragraph (2)(a) must be read in light of Article 10 which generally excludes *renvoi*. Therefore, the expression "law in force in a territorial unit of a Multi-unit State" means the law in force *other than conflict of laws rules*, that is, substantive law rules. In other words, the provision, by itself, cannot produce *renvoi* across international borders (*i.e.,* the further reference to the law of a territorial unit of a different Multi-unit State or the law of a different State). Likewise, Paragraph (2)(a) cannot produce what one might conceive of as an *internal renvoi* (*i.e.,* the further reference to the law of a different territorial unit of that Multi-unit State); this results from its plain language. The only *renvoi*-type mechanisms permitted under the Convention, which might be seen as exceptions from the principle reflected in Article 10, are: (a) the limited form of *internal renvoi* that might arise under Article 12(2)(b) (see paras. 12-14 and 12-15), and (b) the limited form of *internal renvoi* that might arise under Article 5, if a Multi-unit State makes a declaration under Article 12(3) (see paras. 12-21 *et seq.*).

IV. Article 12(2)(b): Preservation of internal conflict of laws rules in respect of perfection by filing, recording or registration

12-14 As already pointed out, Article 12(2)(b) allows a narrowly circumscribed *internal renvoi* for the rules to govern perfection by public filing, recording or registration. This norm provides the only occasion when respect is given to the internal conflict of laws rules within a Multi-unit State in circumstances where the applicable law is determined by Article 4 (*i.e.,* the primary rule of the Convention), or under one of the fall-back rules in Article 5. Paragraph (2)(b) is an important provision that comes into play with respect to several Multi-unit States. It addresses the fact that, under the laws of several Canadian provinces and U.S. states (although the issue might just as well arise in other Multi-unit States), the law governing the perfection of a security interest in securities by public registration or filing is the law of the jurisdiction in which the debtor is located and the registration or filing must be made in that jurisdiction.[32] Paragraph (2)(b) provides that if "the law in force in a territorial unit of a Multi-unit State" (see paras. 12-12 and 12-13) designates the law of another territorial unit of that State to govern perfection by public filing, recording or registration, the

---

[32] For Canada, see, *e.g.*, the Personal Property Security Act, R.S.O. 1990, c. P.10 (OPPSA), s. 7(1)(b); for the United States, see UCC Sections 9-301, 9-305(c)(1) and 9-307.

law of that other territorial unit governs that issue. The provision thus allows for a severely restricted form of *internal renvoi* by effectuating the conflict of laws rules *internal* to Multi-unit States – but *only* those relating to perfection, and then only rules governing perfection by filing, recording or registration. Note that this rule applies by virtue of the Convention's text and does not depend on the existence of a declaration by the Multi-unit State.

Example 12-4

Gotham Broker, an intermediary organised under the laws of the state of Delaware, and its customer, Investor I, who resides in the state of New Jersey, enter into an account agreement which expressly provides that it is governed by the laws of New York and does not expressly provide that a different law governs all the Article 2(1) issues. At the time the governing law agreement was made, Gotham Broker had a Qualifying Office in New York. Later, Investor I grants to Japanese Bank a security interest in the securities account and all securities credited to the account. Article 4(1) would designate the law of New York as the law applicable to all the issues specified in Article 2(1), including the perfection requirements. If, as would usually be the case, Japanese Bank perfected its security interest by taking "control" within the meaning of New York substantive law, New York law would likewise be the applicable law if the Convention did not exist; this is because, under UCC section 9-305(a)(3) (identical in all 50 states), the law governing perfection and priorities when, with respect to security interests in security entitlements and securities accounts, perfection is achieved by a method other than filing, recording or registration is the internal law of the "securities' intermediary's jurisdiction as specified in Section 8-110(e)", which, in this case, is the jurisdiction designated in the express governing law clause in the account agreement. In that event, Article 12(2)(b) is not called into play because it applies only with respect to conflict of laws provisions relating to perfection by filing, recording or registration.

If, however, Japanese Bank chose to perfect its security interest by filing, then, if the Convention did not exist, New Jersey law would govern with respect to perfection by filing of security interests in investment property (a term that would encompass the collateral in this example); this is because under UCC 9-305(c)(1) (identical in all 50 states), the issue of perfection, when achieved by filing, is governed by the internal law of the jurisdiction where the debtor is located (*i.e.*, the same rule as holds true for most other types of collateral perfected by filing). Since Japanese Bank perfected its security interest by filing and New York law designates the law of New Jersey (another state of the same Multi-unit State), Article 12(2)(b) is called into play; it effectuates (rather than disrupts) the New York conflict of laws scheme with respect to perfection by filing, designating the law of New Jersey for purposes of the Convention with respect to the issue of perfection.

12-15 Article 12(2)(b) expressly effectuates only an *internal renvoi*, that is, it provides for the honouring of the designation of applicable substantive law by the conflict of laws rules in force in a territorial unit of a Multi-unit State only of the law of another territorial unit *within the same Multi-unit State* (see the words "of that State" used in the provision). In other words, if, in the example 12-4, the debtor were located (as determined in accordance with UCC Section 9-307(b)) in New Zealand, the designation by New York law, of the law of New Zealand as governing perfection by filing, would not be effectuated by Article 12(2)(b) and New York law itself would apply. The failure of the Convention to include a provision that would effectuate such a transnational cross-border reference is not inadvertent; the preservation of Article 10's stricture against true (*i.e.*, international) *renvoi* was part of the agreed solution reflected in Article 12(2)(b). From the standpoint of domestic practice in the United States, it should be borne in mind that perfection by filing with respect to securities accounts and security entitlements is the exception rather than the rule, and the frequency of a debtor's being located (under UCC Section 9-307) outside the United States also should for the most part be visible (even if not rare); moreover, even in those cases where a "foreign" debtor is involved, the limitation is inapplicable if perfection is by a method other than filing, recording or registration.

V. Article 12(3): Possibility for Multi-unit States to make a declaration preserving internal conflict of laws rules in the context of Article 5

12-16 As is the case under paragraph (2)(b), Article 12(3) allows for a form of *internal renvoi* within Multi-unit States. Under this provision, a Multi-Unit State may file a declaration to the effect that if the applicable law, determined under Article 5, is that of the declaring Multi-unit State itself or that of one of its territorial units, the *internal* conflict of laws rules in force in that Multi-unit State must be applied; consequently, it is those rules which will determine whether the substantive law of that Multi-unit State or of a particular territorial unit of that Multi-unit state shall apply. It is not anticipated that the application of paragraph (3) will often affect outcomes since it comes into play *only* if the applicable law is determined under one of Article 5's fall-back rules and *only* if a Multi-unit State makes a declaration under Article 12(3).

12-17 The second sentence of Article 12(3) imposes deliberately a plain obligation ("shall" not "may") on a Multi-unit State that makes such a declaration to communicate "information concerning the content of those internal choice of law rules" to the Permanent Bureau of the Hague Conference on Private International Law. However, the performance by the declaring State of that obligation is not a condition to the effectiveness of the declaration. A sanction for non-compliance was not felt necessary because providing the information is in the interest of the States concerned. The obligation's purpose is to provide a measure of guidance beyond that provided simply by the existence of the declaration itself, but from the standpoint of functional reliance by a transacting party, it is, of course, understood that "information concerning the content" is not intended to be a substitute for competent specific legal advice. It is anticipated that the information likely would include a copy of the pertinent legislative texts or, if there exists only case law, an explanatory summary of the relevant rules (in the original language and, if possible, also in one of the official languages of the Hague Conference, *i.e.,* in English or French) in force in the Multi-unit State at the time of the declaration. However, if a Multi-unit State does make an Article 12(3) declaration, the obligation under the Convention of all States to apply those rules is not conditioned on the communication or limited to or by the nature or extent of the information communicated.

12-18 The expression "internal choice of law rules", in both sentences of Article 12(3), refers to rules governing choices between the laws of the various territorial units of a Multi-unit State, or between the laws of such territorial units and those of the Multi-unit State itself. Also, the reference is to the internal conflict of laws "rules in force *in*" that Multi-unit State (not "rules *of*" that Multi-unit State). Thus, these internal conflict rules may be either "federal" rules (*i.e.,* rules of the Multi-unit State itself) or rules enacted by a territorial unit designated under Art. 5 (*e.g.*, "state", "provincial" rules etc.). It was clear to the negotiators that there exist multiple models of allocation of competence within Multi-unit States. The focus of paragraph (3) is not to distinguish between provincial or state and national levels of conflict of laws rules; the goal of the provision is to allow a Multi-unit State to preserve its existing internal system, no matter how organised, in the determination of which substantive law (whether federal or territorial unit and, if the latter, which one) should apply; there is no reason why the State should be limited to declaring only the preservation of federal law, and such a limitation was never discussed.

12-19 As is the case under Article 12(2)(b), Article 12(3) does not allow for an *international renvoi* to take place; in other words, the provision cannot lead to the designation of the law of any other State (including another Multi-unit State) or to a territorial unit of another Multi-unit State.

12-20 Finally, in order to ensure absolute predictability of the law applicable, Article 12(3) expressly specifies that the declaration may be made only at the time of signature, ratification, acceptance, approval or accession.

Example 12-5

> Assume that Ruritania has made an Article 12(3) declaration. Investor I subsequently enters into an account agreement with Bank B, an intermediary incorporated under the laws of Ruritania's territorial unit X. The account agreement does not include a choice of law clause, and it does not expressly provide that a particular law governs all the Article 2(1) issues. It does, however, expressly and unambiguously (in a separate paragraph on the signature page of the account agreement) state that the Bank B entered into the account agreement through its Metropolis office in Ruritania's territorial unit Y. Assuming that all conditions of Article 5(1) are satisfied, the applicable law would be the law of the territorial unit Y. However, if a conflict of laws rule of either Ruritania or the territorial unit Y is applicable to these facts and it points to the law, for example, of the place of incorporation of the relevant intermediary, the substantive law of territorial unit X governs all the Article 2(1) issues. Note that the law of territorial unit X governs even if Ruritania has also made an Article 12(4) declaration imposing a geographically more stringent condition. This is because the Article 12(4) declaration has effect only if the applicable law is determined under Article 4. As, in the present case, Article 4 plays no role, the Article 12(4) declaration is irrelevant.

Example 12-6

Assume that the Multi-unit State Pluritania has made an Article 12(3) declaration. Bank B, an intermediary incorporated under the laws of Pluritania itself, and a foreign customer enter into an account agreement. Assume that the law governing all the Article 2(1) issues is determined by Article 5(2). As Bank B is not incorporated under the laws of a territorial unit of Pluritania but instead is incorporated under the laws of Pluritania itself, the applicable law is the law in force in the territorial unit of Pluritania in which Bank has its (principal) place of business. However, if a conflict of laws rule of either Pluritania or the territorial unit of Pluritania in which Bank has its (principal) place of business is applicable to these facts and it points to the law of another territorial unit of Pluritania, the substantive law of that other territorial unit governs all the Article 2(1) issues.

VI. Article 12(4)): Option for a Multi-unit State to impose a geographically more stringent condition

12-21 Article 12(1)(b) provides that the term "that State" in the second sentence of Article 4(1) means the Multi-unit State itself. Thus, in relation to a Multi-unit State, the condition specified in the second sentence of Article 4(1) is satisfied if the relevant intermediary has, at the time of the agreement on governing law, a Qualifying Office anywhere within the Multi-unit State, even if the office is not situated in the territorial unit whose law has been designated by the parties. Article 12(4), however, allows a Multi-unit State to impose, by a declaration, a geographically more stringent condition than the one that would apply under Article 12(1). Of course, if it makes such a declaration, that geographically more stringent condition must be applied by all States where the Convention is in force (not just within the declaring Multi-unit State). The declaration mechanism established by Article 12(4) permits a Contracting State that is a Multi-unit State to declare, at any time, that the condition specified in the second sentence of Article 4(1) is satisfied only if the relevant intermediary has a Qualifying Office within the designated *territorial unit*. If a Multi-unit State makes such a declaration, the parties' agreement on the law of a particular territorial unit of that Multi-unit State will be given effect under Article 4(1) as determinative of the applicable law only if the relevant intermediary has a Qualifying Office within that territorial unit.

12-22 Only a Multi-unit State can make a declaration under Article 12(4) and only with respect to its own territorial units. Thus, a State that is *not* a Multi-unit State cannot make a declaration to the effect that its courts will deem the condition satisfied only if the relevant intermediary had a Qualifying Office within the territorial unit whose law was agreed as the governing law. Also, the making of such a declaration by a Multi-unit State has no effect in any fora (including those within the declaring State) with respect to the designation by the parties of a territorial unit of a Multi-unit State that has *not* made a declaration under Article 12(4).

Example 12-7

Assume that Spain is a Party to the Convention and has made an Article 12(4) declaration. Later, Investor I enters into an account agreement with Bank X, an intermediary organised under the laws of Spain. The account agreement expressly provides that it is governed by the laws of Catalonia and does not expressly provide that a different law governs all the Article 2(1) issues. At the time the governing law agreement was made, Bank X had offices in Madrid, but no office in Catalonia. Still later, Investor I seeks to determine the legal nature and effects against Bank X and third parties of the rights resulting from the credit to its securities account of 100000 shares of Japanese Company. Because Bank X did not, at the time the agreement on governing law was made, have a Qualifying Office in Catalonia, the condition specified in the second sentence of Article 4(1), interpreted in accordance with Article 12(1) but applied in conjunction with the Article 12(4) declaration, is not satisfied. Hence, the applicable law is determined under Article 5.

12-23 Because of the time requirement set forth in Article 4(1), that is, that the relevant intermediary have, *at the time of the agreement on governing law*, an office in the State (*i.e.*, in example 12-7, in Catalonia, as a result of the Art. 12(4) declaration), a conflict of laws clause in an account agreement which is ineffective because the relevant intermediary had no office in the relevant territory when the choice of law clause was agreed upon, remains ineffective even if *later* the relevant intermediary opens an office in the relevant territory (unless the agreement on governing law is subsequently expressly reaffirmed in an amendment to the account agreement and the Qualifying Office test is re-applied and satisfied as of the time of the amendment, see paras. 4-18 and 4-27 *et seq.*).

12-24 Under Article 12(4), a Multi-unit State is not restricted to making a declaration that applies throughout its territory; it may, for example, make a declaration with regard to one or more of its territorial units and, by inaction, allow the general rule of Article 4(1) (as interpreted in accordance with the rules of Art. 12) to apply with respect to its remaining territorial units. For example, Canada could make a declaration that the parties' express designation of the law of Quebec would satisfy the condition specified in the second sentence of Article 4(1) only if the relevant intermediary has a Qualifying Office in Quebec, but allow, by silence as to the other territorial units, the express designation of the law of any of its other provinces to satisfy the condition if the relevant intermediary has a Qualifying Office anywhere in Canada. A particularised declaration applying the special rule to one or several, but less than all, territorial units (if such a declaration is thought by a Multi-unit State to be desirable or even required based on federalism considerations or other internally important grounds) presents no special practical problems. Like all other declarations under the Convention, declarations under Article 12(4) will be made accessible on the Hague Conference's website.

12-25 The last sentence of paragraph 4 provides that a declaration under Article 12(4) shall have no effect on dispositions made *before* the declaration becomes effective. This is important because a declaration under Article 12(4) can be made at any time (unlike a declaration under Art. 12(3)).

Example 12-8

An intermediary and its customer, Investor I, enter into an account agreement which expressly provides that it is governed by the laws of the province of British

Columbia and does not expressly provide that a different law governs all the Article 2(1) issues. At the time the governing law agreement was made, Intermediary had offices only in the provinces of Quebec and Ontario. Later, Investor I grants to Bank B a security interest in the securities account and all securities credited to the account, which Bank B perfects in accordance with the law of British Columbia. Under Article 4(1), as interpreted in accordance with Article 12, all the Article 2(1) issues are governed by the law of British Columbia. Still later, Canada makes a declaration under Article 12(4) with respect to all provinces. Under the last sentence of Article 12(4), the declaration has no impact on dispositions made before the declaration becomes effective, including the grant of the security interest. As a result, even after the Canadian declaration, the security interest remains perfected, but for future dispositions, the pre-declaration agreement on the law of the province of British Columbia is no more effective as long the agreement on the law of the province of British Columbia is not expressly reaffirmed in an amendment to the account agreement and the Qualifying Office test is re-applied and satisfied as of the time of the amendment.

Article 13 *Uniform interpretation*

In the interpretation of this Convention, regard shall be had to its international character and to the need to promote uniformity in its application.

13-1 This Article expresses what has become a standard principle of treaty interpretation (see, *e.g.*, Art. 16 of the *Hague Convention of 1986 on the Law Applicable to Contracts for the International Sale of Goods*, Art. 7(1) of the *United Nations Convention of 1980 on Contracts for the International Sale of Goods*, and Art. 5(1) of the *Cape Town Convention of 2001 on International Interests in Mobile Equipment*). It is an instruction to national courts that they should avoid national canons of interpretation and rather apply an autonomous interpretive regime. Thus, in interpreting the Convention, national courts should take into account not only its wording, using the two authentic texts of the Convention, English and French (which are equally authoritative), and the general purposes of the Convention (in particular, providing legal certainty and predictability, see the first paragraph of the Preamble), but also the interpretation applied by the courts of other Contracting States to it. The purpose of Article 13 is to maintain the maximum degree of uniformity in the interpretation and application of the Convention.

13-2 The importance of this principle is reinforced by the provision in Article 14 for Special Commission meetings to review the practical operation of the Convention (see commentary to Art. 14).

Article 14 ***Review of practical operation of the Convention***

The Secretary General of the Hague Conference on Private International Law shall at regular intervals convene a Special Commission to review the practical operation of this Convention and to consider whether any amendments to this Convention are desirable.

14-1 This Article provides for a mechanism to keep the Convention under review, in particular to consider the judicial interpretation and the application of the terms of the Convention, and to consider whether any modifications to the Convention are desirable in light of evolving practices or problems identified in its interpretation and application. The basic objective of this process is to ensure that the Convention remains effective in providing legal certainty and predictability as to the law applicable to securities credited to a securities account.

14-2 Primary responsibility for convening these meetings lies with the Secretary General of the Hague Conference on Private International Law (HCCH). The Secretary General will normally convene such meetings following a wish expressed by one or more Members of the HCCH, supported by the governing organs of the HCCH. The Secretary General may also take the initiative and propose that such a meeting be held, for example, when the Permanent Bureau (Secretariat), based on the information that it has received or gathered, has found that there is a need for such a meeting.

14-3 All *Members* of the Hague Conference, any other State or Regional Economic Integration Organisation (REIO) *Party* to the Convention, and any *Contracting State* or *Contracting REIO* (on the meaning of the terminology used, see para. 1-43) are invited to attend these review meetings. The Secretary General may also invite other States or International Organisations, be they governmental or non-governmental. As was the case during the preparation of the Convention, adequate private sector industry representation would be sought.

CHAPTER IV TRANSITION PROVISIONS

Article 15 *Priority between pre-Convention and post-Convention interests*

In a Contracting State, the law applicable under this Convention determines whether a person's interest in securities held with an intermediary acquired after this Convention entered into force for that State extinguishes or has priority over another person's interest acquired before this Convention entered into force for that State.

15-1 This Article provides that each Contracting State shall apply the Convention's conflict of laws rule (either Art. 4 or Art. 5) in order to determine the applicable law that governs the issue of priority of interests acquired after the date the Convention enters into force for that Contracting State vis-à-vis interests acquired earlier. These interests are referred to as "post-Convention interests" and "pre-Convention interests" respectively. For the distinction between this Article and Article 7, see the comments in para 7-4. The rule embodied in this Article should cause no difficulty to the parties to a pre-Convention account agreement because they have ample opportunities to adjust their contract by taking account of the effect of this Article. Thus, in many (if not most) situations, the Convention law governing that priority contest will have been anticipated. Nevertheless, in order to guide courts and to enhance certainty for market participants, the Convention deals expressly with this important issue.

15-2 References in this Article to the Convention's entry into force do not refer to its entry into force on the international plane (under Art. 19(1)); rather, they refer to its entry into force for the forum State (whether under Art. 19(1) or Art. 19(2)). The reference at the beginning of this Article to "[i]n a Contracting State" is not intended to have a restricting effect on the applicability of the rule; it simply states the obvious, namely that the rule applies only when the Convention applies (on the meaning of the term "Contracting State", see para. 1-43).

15-3 The rule applies without regard to the nature of the conflicting interests; they might be the same or different types. For example, the priority contest might be between two persons that have security interests in securities credited to a securities account; two account-holders; an account-holder and a person having a security interest in securities credited to a securities account; or a person having a security interest and an attaching creditor.

Example 15-1

> Investor I has securities credited to a securities account maintained for it by an intermediary organised under the laws of Ruritania. Later, Investor I grants to SP-1 a security interest in the securities. SP-1 perfects the security interest in accordance with the substantive law that it believes governs that issue. Still later, the Convention enters into force for Belgium. Still later, Investor I grants to SP-2 a security interest in the securities. SP-2 perfects the security interest in accordance with the substantive law that it believes governs that issue. Eventually, in a proceeding in Belgium, the issue is whether SP-2's security interest has priority

over or extinguished SP-1's security interest. Pursuant to Article 15, the Convention law governs that issue.

15-4 As a result, the application of the Convention law may have an effect on pre-Convention interests (in the above example, the Convention law may determine that the pre-Convention interest is extinguished by the post-Convention interest). Other than this, the Convention remains silent as to whether the Convention law may or must be applied to govern any other issue or be given any other effect as to pre-Convention interests or dispositions (except to the extent otherwise provided in Art. 16).

15-5 It is anticipated that only rarely will priority disputes between pre-Convention dispositions and post-Convention dispositions arise where the entry into force of the Convention will produce a change in the applicable law. It is not the Convention's goal, and rarely will be its effect, to make a change from one settled state of the law to another settled state. The purpose of the Convention is to remove obstacles and uncertainty that under some systems of law stand in the way of the application of the principle embodied in the Convention. Indeed, in several jurisdictions where international clearing systems operate, the applicable law is already determined in accordance with essentially the same rules as those provided in the Convention. Moreover, because of the extensive publicity given to the Convention's rules during their formulation and after the adoption of the Convention, many market participants are already taking the Convention into account. Thus, in many transactions before the Convention has entered into force in a particular State, parties will have anticipated that the Convention's rules might be applicable and will have taken the necessary steps to perfect dispositions also in accordance with the law determined by the Convention.

Article 16 ***Pre-Convention account agreements and securities accounts***

1. **References in this Convention to an account agreement include an account agreement entered into before this Convention entered into force in accordance with Article 19(1). References in this Convention to a securities account include a securities account opened before this Convention entered into force in accordance with Article 19(1).**

2. **Unless an account agreement contains an express reference to this Convention, the courts of a Contracting State shall apply paragraphs (3) and (4) in applying Article 4(1) with respect to account agreements entered into before the entry into force of this Convention for that State in accordance with Article 19. A Contracting State may, at the time of signature, ratification, acceptance, approval or accession, make a declaration that its courts shall not apply those paragraphs with respect to account agreements entered into after the entry into force of this Convention in accordance with Article 19(1) but before the entry into force of this Convention for that State in accordance with Article 19(2). If the Contracting State is a Multi-unit State, it may make such a declaration with respect to any of its territorial units.**

3. **Any express terms of an account agreement which would have the effect, under the rules of the State whose law governs that agreement, that the law in force in a particular State, or a territorial unit of a particular Multi-unit State, applies to any of the issues specified in Article 2(1), shall have the effect that such law governs all the issues specified in Article 2(1), provided that the relevant intermediary had, at the time the agreement was entered into, an office in that State which satisfied the condition specified in the second sentence of Article 4(1). A Contracting State may, at the time of signature, ratification, acceptance, approval or accession, make a declaration that its courts shall not apply this paragraph with respect to an account agreement described in this paragraph in which the parties have expressly agreed that the securities account is maintained in a different State. If the Contracting State is a Multi-unit State, it may make such a declaration with respect to any of its territorial units.**

4. **If the parties to an account agreement, other than an agreement to which paragraph (3) applies, have agreed that the securities account is maintained in a particular State, or a territorial unit of a particular Multi-unit State, the law in force in that State or territorial unit is the law applicable to all the issues specified in Article 2(1), provided that the relevant intermediary had, at the time the agreement was entered into, an office in that State which satisfied the condition specified in the second sentence of Article 4(1). Such an agreement may be express or implied from the terms of the contract considered as a whole or from the surrounding circumstances.**

I. Introduction

16-1 The purpose of this Article is to help market participants avoid costly and unnecessary steps of amending a large number of pre-Convention account agreements or replacing pre-Convention securities accounts in order to enjoy the benefits of the Convention once it comes into force. Article 16 achieves this goal by:

(1) confirming, in Article 16(1), that the Convention applies to account agreements that were entered into, and securities accounts that were opened, before the Convention entered into force on the international plane (Art. 19(1)); and

(2) providing special interpretive rules, in Articles 16(3) and (4), that treat certain provisions of pre-Convention account agreements as having the effect of determining, for the purposes of Article 4(1), the law applicable to all the issues specified in Article 2(1) (the "Article 2(1) issues").

16-2 Article 16(2) directs the courts of each Contracting State to deploy the interpretive rules of Articles 16(3) and (4) in applying Article 4(1) to account agreements entered into before the Convention's entry into force for that State, but only if the account agreement does not contain an express reference to the Convention. If an account agreement contains an express reference to the Convention, Articles 4, 5 and 6 are to be applied directly, without any interpretive assistance from Articles 16(3) and (4).

16-3 Also, Article 16(2) authorises a Contracting State to declare that its courts will not deploy the interpretive rules of Articles 16(3) and (4) in applying Article 4(1) to account agreements entered into during the period between the date the Convention enters into force on the international plane (Art. 19(1)) and the date the Convention enters into force for the declaring State (whether under Art. 19(1) or (2)) (the "Gap Period").

16-4 The interpretive rules of Articles 16(3) and (4) seek to provide legal certainty in the sense that pre-Convention account agreements that include certain specified provisions or terms will be given the proper effect under the Convention.

16-5 For example, existing account agreements governed by state law or federal law in the United States frequently contain provisions either specifying the relevant intermediary's jurisdiction or selecting the governing law (see Example 12-4 and 16-2). These provisions have the effect under rules in the United States of determining the law governing at least some of the issues specified in Article 2(1). Article 16(3) reflects the view that parties who expressly choose such a law to govern their agreement and choose to include one of the specified provisions, may be presumed to have expected the consequences provided by the chosen law and that accordingly it is appropriate to give effect to that expectation.

16-6 Likewise, existing account agreements governed by the law of Member States of the European Community frequently indicate agreement, either expressly or implicitly, on where the relevant securities account is maintained. Article 16(4) reflects the view that parties who have expressly designated the place where the account is maintained may be presumed to have expected the law of that place to govern all the Article 2(1) issues and that accordingly it is appropriate to give effect to that expectation.

16-7 The case may arise where parties have (i) expressly agreed on Location A as the place where the securities account is maintained and (ii) included a provision in the account agreement which selects the law of B as the governing law. Article 16(3) permits a State to opt out of the operation of Article 16(3) in relation to account agreements where the parties have made both of such agreements. Thus, a State which considers that parties would have expected the Article 2(1) issues to be based on the location of the securities account may choose to make the Article 16(3) declaration, thereby ensuring that Article 16(4) will be applicable in all cases where an express location of the account has been specified. Specifically, Article 16(3) authorises each State to declare that its courts will not apply the interpretive rule of Article 16(3) if the parties to the account agreement have expressly agreed that the securities account is maintained in a State *other than* the State whose law would otherwise be applicable by reason of the interpretive rule of Article 16(3). However, if no such declaration is made, Article 16(3) will prevail in determining the applicable law.

16-8 Articles 16(2)-(4) deal only with the interpretation of account agreements; they do not affect Article 4(3), which contains a rule for determining which account agreement is the relevant one for dispositions, from an account holder to its intermediary, of securities held with such intermediary. Article 4(3) applies to both pre- and post-Convention account agreements, whether or not Articles 16(2)-(4) are applicable.

16-9 On the meaning of the term "Contracting State" used in Article 16(2) and (3), see paragraph 1-43.

II. Article 16(1)

16-10 Article 16(1) provides that references in the Convention to "account agreement" and "securities account" include an account agreement entered into prior to the Convention's entry into force on the international plane and include a securities account opened prior to the Convention's entry into force on the international plane (Art. 19(1)). This explicit affirmation that the Convention applies to pre-Convention account agreements and pre-Convention securities accounts eliminates any doubt that might otherwise have existed and prevents any general notion about retroactivity from precluding proper application of the Convention with respect to this particular point.

III. Article 16(2)

A. Providing directions as to the application of Articles 16(3) and (4)

16-11 Article 16(2) directs courts when to apply the interpretive rules of Articles 16(3) and (4). Articles 16(3) and (4) provide rules for interpreting certain pre-Convention account agreements in applying Article 4(1); they do not provide independent conflict of laws rules. These interpretive rules shall be applied to pre-Convention account agreements except in situations where:

(1) the account agreement includes an express reference to the Convention; or

(2) the account agreement is entered into during the Gap Period, and the State in which court proceedings are taking place has made a declaration under

Article 16(2) that its courts will not apply Articles 16(3) or (4) to account agreements entered into during this period.

If any of these two elements is fulfilled, the law governing all the Article 2(1) issues will be determined by direct application of Article 4(1) without any assistance from Articles 16(3) and (4).

16-12 The purpose of limiting the application of Articles 16(3) and (4) to pre-Convention account agreements that do not make express reference to the Convention is to provide informed market participants with a simple method of preventing application of the interpretive rules in Articles 16(3) and (4) to pre-Convention account agreements if they believe those provisions would change the outcome or undermine the legal certainty provided by a direct (*i.e.*, without the interpretive assistance of Arts. 16(3) and (4)) application of Articles 4, 5 and 6. Articles 16(3) and (4) might produce an outcome different from that of direct application of Articles 4, 5 and 6 if the rules of the State whose law governs a pre-Convention account agreement do not treat any of the terms of the account agreement as determining the law applicable to any of the Article 2(1) issues, but rather give that effect (*i.e.*, determining the law applicable to any of the Article 2(1) issues) to terms other than those specified in Articles 4(1) and 5(1), or give a different hierarchical sequence to the terms specified in Articles 4(1) and 5(1). Article 16(4) might be seen as diminishing the legal certainty provided by direct application of Articles 4, 5 and 6 to the extent that it permits courts to consider each account agreement as a whole and the surrounding circumstances. Thus, the "express reference to the Convention" option provides informed market participants with a simple method of having Articles 4, 5 and 6 apply to their pre-Convention account agreements as soon as the Convention enters into force, without exposure to any of the possible different outcomes or uncertainties that might stem from the interpretive assistance of Articles 16(3) and (4). In other words, pre-Convention account agreements which expressly refer to the Convention are excluded from the scope of Article 16 because the parties to such agreements must by definition have tailored their agreements to the terms of the Convention in the same way as the parties to post-Convention agreements. In this circumstance, it is unnecessary and inappropriate to distinguish between pre-Convention and post-Convention agreements.

B. Declaration mechanism

16-13 A State may declare, at the time it signs, ratifies, accepts, approves, or accedes to the Convention, that its courts will not apply Articles 16(3) and (4) to account agreements entered into during the Gap Period between the time the Convention entered into force on the international plane and the time the Convention enters into force for the declaring State. If a Contracting State makes such a declaration, the courts of that State will apply Articles 4, 5 and 6 to Gap Period account agreements without any interpretive assistance from Articles 16(3) or (4). A Multi-Unit State may make such a declaration with respect to any of its territorial units. Declarations are generally covered in Article 20.

Example 16-1

An intermediary and its customer entered into an account agreement in January 2001, before the Convention's entry into force on the international plane. At the time the account agreement was entered into, the intermediary had an office in Zurich that engaged in maintaining securities accounts. The account agreement expressly provides that it is governed by the laws of Switzerland. Under Swiss rules, such a choice of law provision does not have the effect of making Swiss law govern any of the Article 2(1) issues. The intermediary and the account holder want Swiss law to apply to all the Article 2(1) issues as soon as the Convention is entered into force. They cannot rely on the rules of Articles 16(3) or (4) to achieve that result. Article 16(3) will not produce that result because Swiss rules do not treat any terms in any account agreement as determining the law applicable to any of the Article 2(1) issues. Article 16(4) might not produce that result because, although the account agreement does not contain an express agreement as to where the securities account is maintained, it is uncertain whether a court might find an implied agreement from the account agreement as a whole or the surrounding circumstances.

Thus, in January 2003 (*i.e.*, still before the Convention's entry into force on the international plane), the parties amended the account agreement to include an express reference to the Convention. The customer subsequently granted a security interest in all securities credited to the securities account. The secured party perfected its security interest in accordance with Swiss law. Later, after the Convention has come into effect on the international plane, in a proceeding in a State where the Convention is in force, the issue is whether the security interest is perfected. Swiss law governs all the Article 2(1) issues, including the perfection issue. This is because the express reference to the Convention makes Article 4(1) directly applicable without assistance from the interpretive rules of Articles 16(3) or (4); Article 4(1) points to Swiss law because the account agreement contains a governing law clause specifying Swiss law, and the intermediary had, at the time the governing law agreement was made, a Qualifying Office in Switzerland. Article 7 does not apply to this situation as the old law was not determined under the Convention. Article 7 would, however, apply if the parties were to change the account agreement yet another time (either before or after the Convention's entry into force on the international plane) and this new amendment results in a change of the Convention law by application of Article 4(1) (see commentary on Art. 7).

IV. Article 16(3)

16-14 Article 16(3) provides for two matters: (i) an interpretive rule that treats certain provisions of pre-Convention account agreements as having the effect of determining, for purposes of Article 4(1) of the Convention, the law applicable to all the Article 2(1) issues, and (ii) a declaration mechanism.

A. Interpretive rule

16-15 The interpretive rule of Article 16(3) will likely have an impact on pre-Convention account agreements that contain specified provisions and which are governed by the law of certain States, such as the United States, whose conflict of laws rules give choice of law effect to specified provisions in account agreements (see Examples 12-4 and 16-2). Article 16(3) will affect only those pre-Convention account agreements in which any of such provisions is expressly set forth. Article 16(3) provides that if, under the rules of the State or territorial unit of a Multi-Unit State whose law governs the account agreement, those express provisions would have the effect of determining the applicable law for *any* of the Article 2(1) issues, the Convention will have the effect that that law governs *all* the Article 2(1) issues, subject to the Qualifying Office requirement. Where applicable, Article 16(3) can have the effect of expanding the scope of the governing law determination from some to all of the issues specified in Article 2(1), and, in some cases, it might also have the effect of producing a result under Article 4 when otherwise Article 4 would not have produced a result, thereby avoiding recourse to Article 5's fall-back rules. On the meaning of the expression "law in force", see the comments in paragraphs 4-15, 12-12 and 12-13.

B. Declaration mechanism

16-16 A State may, at the time of signature, ratification, acceptance, approval or accession, make a declaration under Article 16(3). The declaration may provide that the courts of that State will not apply the interpretive rule of Article 16(3) if the parties to the account agreement have expressly agreed that the securities account is maintained in a *different* State than the State whose law would otherwise be applicable by reason of the interpretive rule of Article 16(3). A Multi-Unit State may make such a declaration with respect to any of its territorial units.

Example 16-2

> An intermediary incorporated under the laws of Massachusetts and its customer entered into an account agreement in January 2001, before the Convention's entry into force on the international plane. The account agreement states expressly that the securities intermediary's jurisdiction is New York, and expressly provides that the account agreement is governed by the laws of New York. At the time the governing law agreement was made, the intermediary had a Qualifying Office in New York. The customer grants a security interest in the securities account and all security entitlements arising out of a credit of securities to the account. The secured party perfects its security interest in accordance with New York law. Under Section 9-305(a)(3) of the New York Uniform Commercial Code, the selection of the intermediary's jurisdiction has the effect of making New York law applicable to govern the perfection and priority of the security interest, two of the Article 2(1) issues.
>
> Later, in a proceeding in a State where the Convention is in force, the issue is whether the security interest in the securities credited to the account maintained by the intermediary is perfected. The forum State has not made the declaration provided for in Article 16(3) or, if it has, the declaration is irrelevant because the account agreement does not expressly provide that the securities account is maintained in a State other than the United States. New York law would govern all of the Article 2(1) issues, including perfection. This result is determined by Article 4(1) using the interpretive assistance of Article 16(3). The account

agreement's specification of New York as the securities intermediary's jurisdiction has, under New York law, the effect of making New York law applicable to at least one of the Article 2(1) issues (though, in this case, both perfection and priority); the intermediary had, at the time the governing law agreement was made, a Qualifying Office in the United States; and the effect under Article 4(1) of the interpretive assistance of Article 16(3) is to expand the application of New York law to all the Article 2(1) issues.

16-17 If, in Example 16-2, the issue of perfection were litigated in a forum State that had made a declaration under Article 16(3), and the parties had expressly agreed in the account agreement that the securities account is maintained in any State (which need not be the declaring State) other than the State whose law would otherwise have been the applicable law by reason of the interpretive rule of Article 16(3), the court would not apply the interpretive rule of Article 16(3). Instead, the court would apply the interpretive rule of Article 16(4) or, if the applicable law were not determined with the interpretive assistance of Article 16(4), the applicable law would be determined in accordance with Articles 4, 5 and 6 without the assistance of Articles 16(3) or (4).

V. Article 16(4)

16-18 Article 16(4) provides an interpretive rule that treats certain provisions of pre-Convention account agreements as having the effect of determining, for the purposes of Article 4(1) of the Convention, the law applicable to all Article 2(1) issues; it comes into play (i) if Article 16(3) does not apply (see para. 16-15) or (ii) if the applicable law is not determined using the interpretive assistance of Article 16(3) (see paras. 16-16 and 16-17). On the meaning of the expression "law in force", see the comments in paragraphs 4-15, 12-12 and 12-13.

16-19 Thus, if a pre-Convention account agreement to which Article 16(3) is not applicable expressly specifies the State or territorial unit of a Multi-Unit State in which the securities account is maintained, the Convention will treat the law of that State or territorial unit as the law governing the Article 2(1) issues, provided that the relevant intermediary had, at the time the account agreement was entered into, a Qualifying Office in that State.

16-20 In contrast to Article 16(3), under Article 16(4), the agreement as to the State or territorial unit in which the securities account is maintained may be express, implied from the terms of the pre-Convention agreement as a whole, or implied from the surrounding circumstances at the time the parties entered into the pre-Convention agreement. This approach is unique and is found in the Convention only here; it differs sharply from the approach under Article 16(3), which is applicable only if the relevant terms are express, Article 4(1), which requires express agreement, and Article 5(1), which requires that the specified term be "expressly and unambiguously stated in a written account agreement".

16-21 It is important to note that in deciding whether the terms of a pre-Convention agreement considered as a whole or from the surrounding circumstances indicate the existence of an implied agreement as to where the account is maintained, a court is precluded from giving any weight to any of the factors specified in Article 6. Thus, the court may not consider the place where the issuer of the securities is incorporated or otherwise organised, has its statutory seat or registered office, central administration or place or principal place of business; where any certificates representing any underlying securities are located; where the register of underlying securities is located; or where any intermediary other than the relevant intermediary is located. Instead, the court must limit its consideration to factors that are truly relevant to whether the parties reached an implicit agreement on the State or territorial unit in which the securities account is maintained.

Example 16-3

> An intermediary and its customer entered into an account agreement in January 2001, before the Convention's entry into force on the international plane. The account agreement expressly provides that the securities account is maintained in Amsterdam and expressly provides that the account agreement is governed by English law. At the time the account agreement was entered into, the intermediary had a Qualifying Office in the Netherlands. The customer grants a security interest in all securities credited to the securities account. The secured party perfects its security interest in accordance with Dutch law.
>
> Later, in a proceeding in a State where the Convention is in force, the issue is whether the security interest is perfected. The applicable law is not determined by using the interpretive rule of Article 16(3) because English law does not treat any term of an account agreement as having the effect of determining the law applicable to any of the issues in Article 2(1). However, under Article 4(1), using the interpretive rule of Article 16(4), Dutch law governs all the issues specified in Article 2(1), including the issue of perfection. This is because the parties expressly agreed in the account agreement that the securities account is maintained in Amsterdam and the intermediary had, at the time the parties agreed on that, a Qualifying Office in the Netherlands.

Example 16-4

> An intermediary organised under the laws of Ruritania and its customer entered into an account agreement in January 2001, before the Convention's entry into force on the international plane. The account agreement does not contain a governing law clause and does not expressly specify where the securities account is maintained. The account agreement expressly provides, however, that the Ruritanian intermediary entered into the agreement through its office in Metropolis, the capital of Ruritania. At the time the agreement was entered into, the intermediary had a Qualifying Office in Ruritania. The intermediary also assigns a bank code to the customer's securities account that allocates the account to the Metropolis office. The customer's account is credited with securities issued by companies organised in France, England, the United States, Japan, and Australia, which are held by the Ruritanian intermediary through subcustodians in each of those States. The customer grants a security interest in all securities credited to the securities account. The secured party perfects its interest in accordance with Ruritanian law.

Later, in a proceeding in a State where the Convention is in force, the issue is whether the security interest is perfected. For the purpose of ascertaining whether Article 16(3) comes into play, the court (using the conflict of laws rules of the forum, not the Convention) determines that the account agreement is governed by Ruritanian law. The applicable law is not determined by using the interpretive rule of Article 16(3) because Ruritanian law does not treat any term of an account agreement as having the effect of determining the law applicable to any of the Article 2(1) issues. A court could, however, hold, by reason of the interpretive rule of Article 16(4), that Ruritanian law governs all the Article 2(1) issues, including perfection. This is because the clause specifying the Metropolis office as the office through which the Ruritanian intermediary entered into the agreement, and the assignment of the bank code allocating the customer's securities account to the Metropolis office, taken together, could be held to be sufficient to indicate the existence of an implied agreement that the securities account is maintained in Ruritania, and the intermediary had a Qualifying Office in Ruritania at the time the parties agreed on that. The two identified facts may be relied on by the court in finding an implied agreement because they are sufficient to show the intent of the parties to agree as to where the securities account is maintained, and neither fact is among those listed in Article 6.

VI. When applicable law is not determined using Article 16(3) or Article 16(4)

16-22 As noted above, the rules in Articles 16(3) and (4) are only interpretive rules; they do not constitute an independent private international law regime. Thus, if the applicable law is not determined by virtue of the interpretive assistance of those provisions, Articles 4, 5 and 6 determine the applicable law without their assistance. This might occur in a number of circumstances, such as:

- the account agreement was not entered into before the Convention's entry into force on the international plane;
- the account agreement makes express reference to the Convention;
- the pre-Convention account agreement does not contain any of the terms referred to in Articles 16(3) and (4);
- the pre-Convention account agreement was entered into during the Gap Period and the forum is a Contracting State that made a declaration that its courts would not apply Articles 16(3) and (4) to such account agreements;
- the intermediary did not, at the time the account agreement was entered into, have a Qualifying Office in the relevant State or territorial unit; or
- the account agreement does not contain an express or implied agreement as to where the securities account is maintained (and Art. 16(3) also does not apply).

16-23 Applying Articles 4, 5 and 6 under such circumstances without the assistance of the interpretive rules in Articles 16(3) or (4) preserves the overall structure of the Convention: if the applicable law is not determined under Article 4, it will be determined under the fall-back rules of Article 5.

Example 16-5

An intermediary and its customer entered into an account agreement in January 2001, before the Convention's entry into force on the international plane. The account agreement expressly provides that the account agreement is governed by the laws of New Zealand and expressly provides that the securities account is maintained in the Cayman Islands. At the time the agreement was entered into, the intermediary had a Qualifying Office in Auckland. It also had at that time an office in the Cayman Islands, but that office did not satisfy the condition specified in the second sentence of Article 4(1). The customer grants a security interest in all securities credited to the securities account. The secured party perfects its interest in accordance with the laws of New Zealand and the Cayman Islands.

Later, in a proceeding in a State where the Convention is in force, the issue is whether the security interest is perfected. The applicable law is not determined with the assistance of Article 16(3) because under New Zealand law, neither the selection of the governing law nor the agreement of the place where the account is maintained has the effect of determining the law applicable to any of the issues in Article 2(1). The applicable law is not determined with the assistance of Article 16(4) because, although the account agreement included an express agreement that the securities account is maintained in the Cayman Islands, the intermediary did not, at the time the account agreement was entered into, have a Qualifying Office in the Cayman Islands. Thus, Article 4(1) must be applied without the assistance of the interpretive rules of Articles 16(3) and (4). Under Article 4(1), New Zealand law governs all the issues in Article 2(1). This is because the account agreement expressly provides that it is governed by New Zealand law and the intermediary had a Qualifying Office in New Zealand at the time the governing law agreement was made.

Example 16-6

An intermediary and its customer entered into a written account agreement in January 2001, before the Convention's entry into force on the international plane. The account agreement does not contain a governing law clause and does not include an express agreement as to where the securities account is maintained. The account agreement expressly and unambiguously provides, however, that the intermediary entered into the agreement through its Heidelberg office. At the time the agreement was entered into, the intermediary had a Qualifying Office in Heidelberg. The customer grants a security interest in all securities credited to the securities account. The secured party perfects its interest in accordance with German law.

Later, in a proceeding in a State where the Convention is in force, the issue is whether the security interest is perfected. The court determines that the account agreement is governed by German law. The law applicable to the Article 2(1) issues is not determined with the assistance of Article 16(3) because German law does not treat any term of an account agreement as having the effect of determining the law applicable to any of the issues in Article 2(1). The law applicable to the Article 2(1) issues is not determined with the assistance of Article 16(4) because the account agreement did not contain an express agreement as to where the securities account is maintained, and the provision that the agreement was entered into through the Heidelberg office is not by itself sufficient to indicate the existence of an implied agreement as to where the securities account is maintained.

Assume that there is nothing else in the account agreement or in the surrounding circumstances that would support a finding of an implied agreement on that point. The law applicable to all the Article 2(1) issues is not determined under Article 4(1), applied here without the assistance of Articles 16(3) and (4), because the account agreement does not contain an express agreement as to the State whose law governs the account agreement or all the Article 2(1) issues. The fallback rule in Article 5(1) applies (without the assistance of the interpretive rules of Arts. 16(3) and (4), which assist only in the application of Art. 4). Under Article 5(1), German law is the applicable law that governs all the Article 2(1) issues because the account agreement expressly and unambiguously states that the intermediary entered into the agreement through its Heidelberg office and the intermediary had a Qualifying Office in Germany at the time the parties entered into the account agreement.

Example 16-7

An intermediary organised under the laws of Singapore and its customer entered into a written account agreement in January 2001, before the Convention's entry into force on the international plane. The account agreement does not contain a governing law clause, does not expressly specify where the securities account is maintained, and does not expressly and unambiguously state that the intermediary entered into the agreement through a particular office. The customer grants a security interest in all securities credited to the securities account. The secured party perfects its security interest in accordance with Singapore law.

Later, in a proceeding in a State where the Convention is in force, the issue is whether the security interest is perfected. The court determines that the account agreement is governed by Singapore law. For the reasons given in the preceding examples, the law applicable to all the Article 2(1) issues is not determined by Article 4 (neither with nor without the assistance of Arts. 16(3) and (4)). The applicable law also is not determined under Article 5(1), because the account agreement, although written, does not expressly and unambiguously contain a statement that the intermediary entered into the account agreement through a particular office. Under Article 5(2), Singapore law governs all the Article 2(1) issues because the intermediary was organised under Singapore law at the time the parties entered into the agreement.

CHAPTER V FINAL CLAUSES

Article 17 *Signature, ratification, acceptance, approval or accession*

1. This Convention shall be open for signature by all States.

2. This Convention is subject to ratification, acceptance or approval by the signatory States.

3. Any State which does not sign this Convention may accede to it at any time.

4. The instruments of ratification, acceptance, approval or accession shall be deposited with the Ministry of Foreign Affairs of the Kingdom of the Netherlands, Depositary of this Convention.

17-1 This Article provides two methods by which a State may become a Contracting State to the Convention: either by *signing* the Convention and then depositing its instrument of *ratification, acceptance* or *approval* of the Convention with the Depositary (paras. (1), (2) and (4)), or, alternatively, by depositing its instrument of *accession* to the Convention with the Depositary (paras. (3) and (4)). The signing of the Convention obliges the State to "refrain from acts which would defeat the object and purpose" of the Convention (see Art. 18 of the Vienna Convention of 1969 on the Law of Treaties) as long as it is a State signatory to the Convention; the deposit of an instrument of ratification, acceptance, approval or accession constitutes, in each case, an international act whereby a State on the international plane is bound by the treaty (see Art. 2(1)(b) of the Vienna Convention).

17-2 There is no difference in quality or effect between the two methods provided for in this Article. Both methods are available equally to Member States and non-Member States of the Hague Conference on Private International Law. Also, the provision makes no distinction between States that participated in the Diplomatic Conference at which the text of the Convention was adopted and those that did not. This is in contrast to most previous Hague Conventions, which generally provide that only Member States of the Conference can sign and ratify, accept or approve the Convention, whereas non-Member States can only accede to it. Additionally, contrary to most previous Hague Conventions, there is no requirement that an accession be accepted. With a view to offering to the States the broadest possible set of options and flexibility for becoming Contracting States, these differentiations have been eliminated intentionally (following Art. 25 of the *Hague Convention of 22 December 1986 on the Law Applicable to Contracts for the International Sales of Goods*).

17-3 This Article does not impose any time requirement or any other prerequisite for signing, ratifying, accepting, approving or acceding to the Convention. In particular, when a State has signed the Convention, this Article does not impose a time limit for ratification, acceptance or approval; the instrument of ratification, acceptance or approval may even be deposited at the time of signing. The effect of paragraph (3) is that non-signatory States may accede before (as well as after) the Convention enters into force on the international plane under Article 19(1).

17-4 The instruments of ratification, acceptance, approval or accession are deposited with the Depositary of the Convention, that is, the Ministry of Foreign Affairs of the Kingdom of the Netherlands (para. (4)); the Depositary then notifies those indicated in Article 24 of any signature, ratification, acceptance, approval or accession under this Article (see further commentary to Art. 24).

17-5 The Convention's entry into force, both on the international plane and for the particular Contracting State, is governed by Article 19, not by this Article.

17-6 Signature, ratification, acceptance, approval or accession under paragraphs (1)-(3) apply only to States; for Regional Economic Integration Organisations, a corresponding provision is found in Article 18.

Article 18 ***Regional Economic Integration Organisations***

1. **A Regional Economic Integration Organisation which is constituted by sovereign States and has competence over certain matters governed by this Convention may similarly sign, accept, approve or accede to this Convention. The Regional Economic Integration Organisation shall in that case have the rights and obligations of a Contracting State, to the extent that that Organisation has competence over matters governed by this Convention. Where the number of Contracting States is relevant in this Convention, the Regional Economic Integration Organisation shall not count as a Contracting State in addition to its Member States which are Contracting States.**
2. **The Regional Economic Integration Organisation shall, at the time of signature, acceptance, approval or accession, notify the Depositary in writing specifying the matters governed by this Convention in respect of which competence has been transferred to that Organisation by its Member States. The Regional Economic Integration Organisation shall promptly notify the Depositary in writing of any changes to the distribution of competence specified in the notice in accordance with this paragraph and any new transfer of competence.**
3. **Any reference to a "Contracting State" or "Contracting States" in this Convention applies equally to a Regional Economic Integration Organisation where the context so requires.**

18-1 This Article enables each Regional Economic Integration Organisation (REIO) constituted by sovereign States (*e.g.*, the European Community) to sign, accept, approve or accede to the Convention (the absence of the term *ratify* is intentional, as only States ratify Conventions), but only to the extent that it has competence over matters covered by the Convention. The European Community, for example, has adopted several legal instruments that deal with matters covered by this Convention (*Directive 98/26/EC of the European Parliament and of the Council of 19 May 1998 on settlement finality in payment and securities settlement systems*, OJ L 166/45 of 11.6.1998; *Directive 2001/24/EC of the European Parliament and of the Council of 4 April 2001 on the reorganisation and winding up of credit institutions*, OJ L 125/15 of 5.5.2001; and *Directive 2002/47/EC of the European Parliament and of the Council of 6 June 2002 on financial collateral arrangements*, OJ L 168/43 of 27.6.2002). Only the Community, and not its Member States, has competence to conclude international agreements that affect those instruments. For this reason (and because the European Community is not a Multi-unit State within the meaning of the Convention), it is necessary to include a provision in the Convention, permitting the European Community (and any other REIO) to become a Contracting body by providing it with the rights and obligations of a Contracting State (thus becoming a "Contracting REIO"). While this is the first appearance of such a provision in a Hague Convention, this Article closely follows

Article 48 of the *Cape Town Convention of 2001 on International Interests in Mobile Equipment.*[33]

18-2 In view of the importance of this matter, REIOs are to notify the Depositary (*i.e.*, the Ministry of Foreign Affairs of the Kingdom of the Netherlands) in writing (see Art. 1(1)(n)), specifying the matters covered by the Convention in respect of which "competence has been transferred to that Organisation by its Member States". Thus, the notification should be made only where, as a result of the transfer of competence, the REIO has exclusive competence in relation to the specified matters and Member States no longer have independent authority to legislate concerning them. The notification has to be made at the time of signature or of the deposit of the instrument of acceptance, approval or accession; REIOs must "promptly" notify the Depositary of all changes, if any, to the distribution of competence and all new transfers, if any, of competence. These notifications under Article 18(2) are *not* to be considered as declarations covered by Article 22: Notifications under Article 18 are compulsory, whereas declarations under Article 22 are not.

18-3 A Contracting REIO has, within the limits of its competence in these matters (para. (1) of this Article), the same rights and duties as a Contracting State. Thus, paragraph (3) of this Article provides that any provisions concerning a Contracting State in the Convention apply equally to a Contracting REIO where the context so requires. The Convention, however, specifies that where the number of Contracting States is relevant in this Convention, the Contracting REIO shall not count as a Contracting State in addition to its Member States which are Contracting States" (para. (1), third sentence). This is relevant for the Convention's entry into force on the international plane (Art. 19(1)), which requires the deposit of instruments of ratification, acceptance, approval or accession by three *States*. Thus, the deposit of an instrument of acceptance, approval or accession by an REIO does not count for the purpose of Article 19(1).

18-4 The instrument of acceptance, approval or accession by an REIO shall be deposited with the Depositary. This is not expressly stated in the Convention, but follows by analogy from Article 17(4); this is because the word "similarly" in the first sentence of paragraph (1) suggests that Article 17(4), which only deals with treaty-administering aspects, also applies in the context of REIOs. Note that under Article 24 the Depositary notifies all Members of the Hague Conference on Private International Law, as well as all other States and REIOs which have already signed the Convention or deposited their instrument(s) of ratification, acceptance, approval or accession, of all signatures and all deposits of instruments of acceptance, approval or accession of an REIO under paragraph (1) of this Article, and of all notifications made under paragraph (2) of this Article (see Arts. 24(a) and (d)).

---

33 The comments on this Article, therefore, closely follow the official comments on Art. 48 of the Cape Town Convention, see *Convention on International Interests in Mobile Equipment and Protocol Thereto on Matters Specific to Aircraft Equipment, Official Commentary by Professor Sir Roy Goode CBE, QC*, published and distributed by the International Institute for the Unification of Private Law (UNIDROIT), Rome, September 2002, pp. 150-152.

Article 19 ***Entry into force***

1. This Convention shall enter into force on the first day of the month following the expiration of three months after the deposit of the third instrument of ratification, acceptance, approval or accession referred to in Article 17.

2. Thereafter this Convention shall enter into force –

a) for each State or Regional Economic Integration Organisation referred to in Article 18 subsequently ratifying, accepting, approving or acceding to it, on the first day of the month following the expiration of three months after the deposit of its instrument of ratification, acceptance, approval or accession;

b) for a territorial unit to which this Convention has been extended in accordance with Article 20(1), on the first day of the month following the expiration of three months after the notification of the declaration referred to in that Article.

I. Introduction

19-1 Article 19 addresses two matters relating to the entry into force of the Convention. First, it determines when the Convention enters into force on the international plane (Art. 19(1)). Secondly, it determines when the Convention enters into force for a particular State (including territorial units (of a Multi-unit State) to which the Convention has been extended) or a Regional Economic Integration Organisation (REIO) that has deposited its instrument of ratification, acceptance, approval or accession. With respect to this second matter, Article 19 further distinguishes between the States that bring the Convention into force on the international plane (Art. 19(1)), on the one hand, and any subsequent State or any REIO (Art. 19(2)) on the other.

19-2 The Depositary notifies of each date of an entry into force of the Convention (under either Art. 19(1) or (2); Art. 24(b)).

19-3 The mere signing of the Convention without the deposit of an instrument of ratification, acceptance, or approval cannot lead to the Convention's entry into force (neither under Art. 19(1) nor (2)).

II. Entry into force of the Convention on the international plane (Art. 19(1))

19-4 The Convention enters into force on the international plane on the first day of the month following the expiration of three months after the deposit of the third instrument of ratification, acceptance, approval or accession by a State. Thus, if, for example, the third instrument is deposited on July 18th, the Convention enters into force on November 1st. This is true even if November 1st is a Saturday, Sunday or holiday. The deposit of an instrument by a REIO under Article 18 does not count under Article 19(1); only instruments deposited by a State are taken into account for this purpose.

19-5 The date of entry into force of the Convention on the international plane has significance for the operation of Article 16. That date is the dividing line between pre- and post-Convention account agreements and securities accounts; also it is the start-date of the "Gap-Period" (relevant in the event of a declaration under Art. 16(2)). Because the Convention's entry into force on the international plane is based on the date of deposits with the Depositary, the location of the Depositary determines the relevant time zone.

III. Entry into force of the Convention for a State, Regional Economic Integration Organisation or a territorial unit

A. For the States that bring the Convention into force (Art. 19(1))

19-6 For the three States that bring the Convention into force on the international plane, the Convention enters into force on the same day. This result is not expressly stated in Article 19 because it is self-evident; it would be contrary to established treaty practice for the Convention to enter into force on the international plane, but not to be in force for the States that made it enter into force.

19-7 When does the Convention enter into force for a territorial unit of a Multi-unit State that is among the first three States? On the same date as it enters into force for the State itself. Depending on the facts, this result is reached by either of two paths. Under Article 20(3), if a State makes no declaration under Article 20(1), the Convention by its terms extends to all the territorial units of that State. On the other hand, under Article 22(d), if a Multi-unit State makes a declaration under Article 20(1) (which extends the Convention to one or more of its territorial units and which may be made only at the time of signature of the Convention or at the time of the deposit of the instrument of ratification, acceptance, approval or accession), that declaration takes effect simultaneously with the entry into force of the Convention for the State concerned.

B. For subsequent States and for REIOs (Art. 19(2))

19-8 Article 19(2) addresses the entry into force for the fourth and following States and for REIOs. For these, the Convention enters into force on the first day of the month following the expiration of three months after the deposit of their respective instrument (Art. 19(2)(a)). The word "subsequently" in Article 19(2)(a) does not refer to the time following the entry into force of the Convention on the international plane, but rather to the time following the deposit of the third instrument of ratification, acceptance, approval or accession. The deposit of the fourth instrument might even occur on the same day as the deposit of the third, and in that event the Convention would enter into force for the first three States and the fourth depositor on the same day.

19-9 When does the Convention enter into force for a territorial unit of a Multi-unit State that is not among the first three depositing States? On the same date as it enters into force for the State itself. Depending on the facts, this result is reached by either of two paths. Under Article 20(3), if a State makes no declaration under Article 20(1), the Convention by its terms extends to all the territorial units of that State (the same rule as for the first three depositing States). On the other hand, if a Multi-unit State makes a declaration under Article 20(1), this result is reached by Article 19(2)(b) if the declaration is notified concurrently with the deposit of the instrument of ratification, acceptance, approval or accession. It is therefore highly advisable for a Multi-unit State that wishes to make an extension declaration to notify that declaration concurrently with the deposit of its instrument of ratification, acceptance, approval or accession.

Article 20 ***Multi-unit States***

1. **A Multi-unit State may, at the time of signature, ratification, acceptance, approval or accession, make a declaration that this Convention shall extend to all its territorial units or only to one or more of them.**
2. **Any such declaration shall state expressly the territorial units to which this Convention applies.**
3. **If a State makes no declaration under paragraph (1), this Convention extends to all territorial units of that State.**

20-1 Article 20 provides what is sometimes loosely referred to as a federal State extension clause. It is applicable, however, to all Multi-unit States as defined in Art. 1(1)(m) and is not limited to those based on a "federal" relationship. It is a clause typically found in modern (Hague) Conventions. It provides that a Multi-unit State may declare, at the time of signature or at the time of depositing its instrument of ratification, acceptance, approval or accession, that the Convention shall extend to all its territorial units or only to one or more of them (Art. 20(1)). If it does not make a declaration, the Convention applies to all the territorial units.

20-2 A declaration made under Article 20(1) can be amended at any time by submitting a new declaration under Article 22(b) or can be withdrawn at any time under Article 22(c). If the declaration is withdrawn, the Convention will extend to all territorial units of that Multi-unit State; in such a case, the Convention enters into force for the territorial units in which the Convention had not been in force previously, six months after the date on which the Depositary made the notification under Article 24 (see Art. 22(e)). If, however, the declaration is amended by submitting a new declaration that extends the Convention to one or more additional territorial units, the Convention enters into force for these additional territorial unit(s) on the first day of the month following the expiration of three months after the date on which the Depositary made the notification in accordance with Article 24 (see Art. 22(d), second provision). In both cases, the date of entry into force of the Convention for the territorial units in which the Convention had already been in force does not change.

20-3 If the law determined by the Convention (whether under Art. 4 or 5) is the law of a territorial unit to which the Convention has not been extended, the law of that territorial unit may nonetheless be applied. This result is reached by analogy with Article 9, which states that the Convention is of general applicability and hence will apply whether or not the applicable law is that of a Contracting State.

Article 21 *Reservations*

No reservation to this Convention shall be permitted.

21-1 Like most other Hague Conventions, the Convention does not allow for any reservation. A reservation is a unilateral act by which a State purports to exclude or modify the legal effect of a particular provision of a treaty in its application to that State (see Art. 2(1)(d) of the Vienna Convention on the Law of Treaties). Under modern public international law, unless expressly authorised by the Treaty, a reservation is not binding on other States unless they accept it (Art. 20 of the Vienna Convention on the Law of Treaties). Hague Conventions are traditionally hostile to reservations; only a few Hague Conventions allow for some expressly specified reservations. In the case of the Securities Convention, no Member State voiced the need for a specific reservation. As a result, no reservation clause was embodied in the Convention and thus no reservation is permitted.

21-2 The Convention does allow for certain declarations (Art. 22); these are not to be regarded as reservations. These declarations involve adjustments within the limit of the terms permitted by the Convention itself and apply only with respect to the territory of the State making the declaration.

Article 22 ***Declarations***

For the purposes of Articles 1(5), 12(3) and (4), 16(2) and (3) and 20 –

a) any declaration shall be notified in writing to the Depositary;

b) any Contracting State may modify a declaration by submitting a new declaration at any time;

c) any Contracting State may withdraw a declaration at any time;

d) any declaration made at the time of signature, ratification, acceptance, approval or accession shall take effect simultaneously with the entry into force of this Convention for the State concerned; any declaration made at a subsequent time and any new declaration shall take effect on the first day of the month following the expiration of three months after the date on which the Depositary made the notification in accordance with Article 24;

e) a withdrawal of a declaration shall take effect on the first day of the month following the expiration of six months after the date on which the Depositary made the notification in accordance with Article 24.

22-1 Article 22 provides the technical and administrative rules governing the manner in which the declarations permitted under the Convention may be made, modified and withdrawn, and when they become effective. This provision refers and applies to all six possibilities for declaration provided under the Convention (see Arts. 1(5), 12(3) and (4), 16(2) and (3), and 20).

22-2 Article 22(a) provides that declarations shall be notified in writing to the Depositary (see para. 24-1). Under the definition in Article 1(1)(n) of the term "writing", a declaration may be sent to the Depositary in electronic form, provided that the declaration is capable of being reproduced in tangible form on a subsequent occasion. Working in conjunction with the Depositary, the Permanent Bureau of the Hague Conference will provide timely notice on the HCCH's website of the declarations made in conformity with Article 22.

22-3 Pursuant to Article 22(b), a declaration may not be modified by amending the existing version of the declaration, but rather by submitting a new declaration. This may be done at any time. This methodology facilitates access to updated information, as anyone wishing to ascertain the current content of a declaration will be able to do so by consulting and relying on one complete, free-standing record (thus not having to piece together an original declaration and one or more amendments).

22-4 Article 22(c) states that a declaration may be withdrawn at any time. This is done by submitting a new declaration stating the withdrawal. The withdrawal of a declaration takes effect on the first day of the month following the expiration of six months after the date on which the Depositary made the notification in accordance with Article 24. This longer period takes into account that withdrawals may require a longer period of adjustment for the marketplace.

22-5 Article 22(d) addresses the question as to when a declaration shall take effect. Article 22(d) distinguishes two situations, based on when a declaration is made. If a declaration is made at the time of signature, or at the time of depositing an instrument of ratification, acceptance, approval or accession (this may be a declaration under Arts. 1(5), 12(3), 12(4), 16(2), 16(3), or 20), it will take effect simultaneously with the entry into force of the Convention (whether under Art. 19(1) or (2)) for the State concerned. If, however, a declaration is made at a subsequent time (this may be an original declaration under Art. 1(5) or 12(4) or a new declaration amending an earlier declaration), it will take effect on the first day of the month following the expiration of three months after the date on which the Depositary makes the notification pursuant to Article 24 (see commentary to Art. 24). This is a slight departure from existing practice, as generally in Hague Conventions, the period leading to the date when a declaration takes effect starts running when the Depository receives the notification from the State making the declaration. In this Convention, however, it seemed more appropriate to let the time period begin to run once the notification has been sent out by the Depositary. This change in system approach was introduced to ensure that the marketplace has the necessary time to adjust to the effects of a declaration.

Article 23 ***Denunciation***

1. **A Contracting State may denounce this Convention by a notification in writing to the Depositary. The denunciation may be limited to certain territorial units of a Multi-unit State to which this Convention applies.**
2. **The denunciation shall take effect on the first day of the month following the expiration of twelve months after the date on which the notification is received by the Depositary. Where a longer period for the denunciation to take effect is specified in the notification, the denunciation shall take effect upon the expiration of such longer period after the date on which the notification is received by the Depositary.**

23-1 The duration of the Convention is unlimited and thus can come to an end only by means of a denunciation. Article 23 permits a Contracting State to withdraw from the Convention for itself or any of its territorial units by notification in writing to the Depositary. Such a denunciation will become effective either on the first day of the month following the expiration of twelve months after the date on which the notification was received by the Depositary, or, if the notification specifies a longer period, upon the expiration of that longer period after the date on which the notification was received by the Depositary.

23-2 The Depositary notifies all Members of the Hague Conference on Private International Law, as well as any other State or REIO which has already signed, ratified, accepted, approved or acceded to the Convention, of the date of any denunciation made under Article 23 (see Art. 24(e)).

Article 24 ***Notifications by the Depositary***

The Depositary shall notify the Members of the Hague Conference on Private International Law, and other States and Regional Economic Integration Organisations which have signed, ratified, accepted, approved or acceded in accordance with Articles 17 and 18, of the following –

a) the signatures and ratifications, acceptances, approvals and accessions referred to in Articles 17 and 18;

b) the date on which this Convention enters into force in accordance with Article 19;

c) the declarations and withdrawals of declarations referred to in Article 22;

d) the notifications referred to in Article 18(2);

e) the denunciations referred to in Article 23.

24-1 This traditional provision lists the responsibilities of the Treaty Depositary, that is, the Ministry of Foreign Affairs of the Kingdom of the Netherlands. These responsibilities have already been referred to in the comments on the relevant provisions.

24-2 In addition to the responsibilities expressly listed in Article 24, the Treaty Depositary will also perform such other functions as are customary for treaty depositaries. These include keeping custody of the original text of the Convention and of instruments of ratification, acceptance, approval or accession, and declarations; receiving signatures of the Convention; examining each signature and each instrument to ensure that it is in the proper form; and registering the Convention with the Secretariat of the United Nations upon its coming into force (Art. 77(1) of the Vienna Convention of 1969 on the Law of Treaties). The Treaty Depositary is also in a position to give guidance on such matters as the standards it employs to determine the acceptability of instruments and the format of declarations to be deposited with it.

PART III:
APPENDIXES

APPENDIX 1

TEXT OF THE HAGUE SECURITIES CONVENTION

Convention on the Law Applicable to Certain Rights in Respect of Securities held with an Intermediary

The States signatory to the present Convention,

Aware of the urgent practical need in a large and growing global financial market to provide legal certainty and predictability as to the law applicable to securities that are now commonly held through clearing and settlement systems or other intermediaries,

Conscious of the importance of reducing legal risk, systemic risk and associated costs in relation to cross-border transactions involving securities held with an intermediary so as to facilitate the international flow of capital and access to capital markets,

Desiring to establish common provisions on the law applicable to securities held with an intermediary beneficial to States at all levels of economic development,

Recognising that the "Place of the Relevant Intermediary Approach" (or PRIMA) as determined by account agreements with intermediaries provides the necessary legal certainty and predictability,

Have resolved to conclude a Convention to this effect, and have agreed upon the following provisions –

CHAPTER I – DEFINITIONS AND SCOPE OF APPLICATION

Article 1 *Definitions and interpretation*

1. In this Convention –

   a) "securities" means any shares, bonds or other financial instruments or financial assets (other than cash), or any interest therein;

   b) "securities account" means an account maintained by an intermediary to which securities may be credited or debited;

   c) "intermediary" means a person that in the course of a business or other regular activity maintains securities accounts for others or both for others and for its own account and is acting in that capacity;

   d) "account holder" means a person in whose name an intermediary maintains a securities account;

   e) "account agreement" means, in relation to a securities account, the agreement with the relevant intermediary governing that securities account;

   f) "securities held with an intermediary" means the rights of an account holder resulting from a credit of securities to a securities account;

g) "relevant intermediary" means the intermediary that maintains the securities account for the account holder;

h) "disposition" means any transfer of title whether outright or by way of security and any grant of a security interest, whether possessory or non-possessory;

i) "perfection" means completion of any steps necessary to render a disposition effective against persons who are not parties to that disposition;

j) "office" means, in relation to an intermediary, a place of business at which any of the activities of the intermediary are carried on, excluding a place of business which is intended to be merely temporary and a place of business of any person other than the intermediary;

k) "insolvency proceeding" means a collective judicial or administrative proceeding, including an interim proceeding, in which the assets and affairs of the debtor are subject to control or supervision by a court or other competent authority for the purpose of reorganisation or liquidation;

l) "insolvency administrator" means a person authorised to administer a reorganisation or liquidation, including one authorised on an interim basis, and includes a debtor in possession if permitted by the applicable insolvency law;

m) "Multi-unit State" means a State within which two or more territorial units of that State, or both the State and one or more of its territorial units, have their own rules of law in respect of any of the issues specified in Article 2(1);

n) "writing" and "written" mean a record of information (including information communicated by teletransmission) which is in tangible or other form and is capable of being reproduced in tangible form on a subsequent occasion.

2. References in this Convention to a disposition of securities held with an intermediary include –

a) a disposition of a securities account;

b) a disposition in favour of the account holder's intermediary;

c) a lien by operation of law in favour of the account holder's intermediary in respect of any claim arising in connection with the maintenance and operation of a securities account.

3. A person shall not be considered an intermediary for the purposes of this Convention merely because –

a) it acts as registrar or transfer agent for an issuer of securities; or

b) it records in its own books details of securities credited to securities accounts maintained by an intermediary in the names of other persons for whom it acts as manager or agent or otherwise in a purely administrative capacity.

4. Subject to paragraph (5), a person shall be regarded as an intermediary for the purposes of this Convention in relation to securities which are credited to securities accounts which it maintains in the capacity of a central securities depository or which

are otherwise transferable by book entry across securities accounts which it maintains.

5. In relation to securities which are credited to securities accounts maintained by a person in the capacity of operator of a system for the holding and transfer of such securities on records of the issuer or other records which constitute the primary record of entitlement to them as against the issuer, the Contracting State under whose law those securities are constituted may, at any time, make a declaration that the person which operates that system shall not be an intermediary for the purposes of this Convention.

Article 2 Scope of the Convention and of the applicable law

1. This Convention determines the law applicable to the following issues in respect of securities held with an intermediary –
   a) the legal nature and effects against the intermediary and third parties of the rights resulting from a credit of securities to a securities account;
   b) the legal nature and effects against the intermediary and third parties of a disposition of securities held with an intermediary;
   c) the requirements, if any, for perfection of a disposition of securities held with an intermediary;
   d) whether a person's interest in securities held with an intermediary extinguishes or has priority over another person's interest;
   e) the duties, if any, of an intermediary to a person other than the account holder who asserts in competition with the account holder or another person an interest in securities held with that intermediary;
   f) the requirements, if any, for the realisation of an interest in securities held with an intermediary;
   g) whether a disposition of securities held with an intermediary extends to entitlements to dividends, income, or other distributions, or to redemption, sale or other proceeds.
2. This Convention determines the law applicable to the issues specified in paragraph (1) in relation to a disposition of or an interest in securities held with an intermediary even if the rights resulting from the credit of those securities to a securities account are determined in accordance with paragraph (1)(a) to be contractual in nature.
3. Subject to paragraph (2), this Convention does not determine the law applicable to –
   a) the rights and duties arising from the credit of securities to a securities account to the extent that such rights or duties are purely contractual or otherwise purely personal;
   b) the contractual or other personal rights and duties of parties to a disposition of securities held with an intermediary; or

c) the rights and duties of an issuer of securities or of an issuer's registrar or transfer agent, whether in relation to the holder of the securities or any other person.

Article 3 Internationality

This Convention applies in all cases involving a choice between the laws of different States.

CHAPTER II –APPLICABLE LAW

Article 4 Primary rule

1. The law applicable to all the issues specified in Article 2(1) is the law in force in the State expressly agreed in the account agreement as the State whose law governs the account agreement or, if the account agreement expressly provides that another law is applicable to all such issues, that other law. The law designated in accordance with this provision applies only if the relevant intermediary has, at the time of the agreement, an office in that State, which –

   a) alone or together with other offices of the relevant intermediary or with other persons acting for the relevant intermediary in that or another State –

      i) effects or monitors entries to securities accounts;

      ii) administers payments or corporate actions relating to securities held with the intermediary; or

      iii) is otherwise engaged in a business or other regular activity of maintaining securities accounts; or

   b) is identified by an account number, bank code, or other specific means of identification as maintaining securities accounts in that State.

2. For the purposes of paragraph (1)(a), an office is not engaged in a business or other regular activity of maintaining securities accounts –

   a) merely because it is a place where the technology supporting the bookkeeping or data processing for securities accounts is located;

   b) merely because it is a place where call centres for communication with account holders are located or operated;

   c) merely because it is a place where the mailing relating to securities accounts is organised or files or archives are located; or

   d) if it engages solely in representational functions or administrative functions, other than those related to the opening or maintenance of securities accounts, and does not have authority to make any binding decision to enter into any account agreement.

3. In relation to a disposition by an account holder of securities held with a particular intermediary in favour of that intermediary, whether or not that intermediary

maintains a securities account on its own records for which it is the account holder, for the purposes of this Convention –

a) that intermediary is the relevant intermediary;

b) the account agreement between the account holder and that intermediary is the relevant account agreement;

c) the securities account for the purposes of Article 5(2) and (3) is the securities account to which the securities are credited immediately before the disposition.

Article 5 *Fall-back rules*

1. If the applicable law is not determined under Article 4, but it is expressly and unambiguously stated in a written account agreement that the relevant intermediary entered into the account agreement through a particular office, the law applicable to all the issues specified in Article 2(1) is the law in force in the State, or the territorial unit of a Multi-unit State, in which that office was then located, provided that such office then satisfied the condition specified in the second sentence of Article 4(1). In determining whether an account agreement expressly and unambiguously states that the relevant intermediary entered into the account agreement through a particular office, none of the following shall be considered –

   a) a provision that notices or other documents shall or may be served on the relevant intermediary at that office;

   b) a provision that legal proceedings shall or may be instituted against the relevant intermediary in a particular State or in a particular territorial unit of a Multi-unit State;

   c) a provision that any statement or other document shall or may be provided by the relevant intermediary from that office;

   d) a provision that any service shall or may be provided by the relevant intermediary from that office;

   e) a provision that any operation or function shall or may be carried on or performed by the relevant intermediary at that office.

2. If the applicable law is not determined under paragraph (1), that law is the law in force in the State, or the territorial unit of a Multi-unit State, under whose law the relevant intermediary is incorporated or otherwise organised at the time the written account agreement is entered into or, if there is no such agreement, at the time the securities account was opened; if, however, the relevant intermediary is incorporated or otherwise organised under the law of a Multi-unit State and not that of one of its territorial units, the applicable law is the law in force in the territorial unit of that Multi-unit State in which the relevant intermediary has its place of business, or, if the relevant intermediary has more than one place of business, its principal place of business, at the time the written account agreement is entered into or, if there is no such agreement, at the time the securities account was opened.

3. If the applicable law is not determined under either paragraph (1) or paragraph (2), that law is the law in force in the State, or the territorial unit of a Multi-unit State, in which the relevant intermediary has its place of business, or, if the relevant intermediary has more than one place of business, its principal place of business, at the time the written account agreement is entered into or, if there is no such agreement, at the time the securities account was opened.

Article 6 Factors to be disregarded

In determining the applicable law in accordance with this Convention, no account shall be taken of the following factors –

a) the place where the issuer of the securities is incorporated or otherwise organised or has its statutory seat or registered office, central administration or place or principal place of business;

b) the places where certificates representing or evidencing securities are located;

c) the place where a register of holders of securities maintained by or on behalf of the issuer of the securities is located; or

d) the place where any intermediary other than the relevant intermediary is located.

Article 7 Protection of rights on change of the applicable law

1. This Article applies if an account agreement is amended so as to change the applicable law under this Convention.

2. In this Article –

a) "the new law" means the law applicable under this Convention after the change;

b) "the old law" means the law applicable under this Convention before the change.

3. Subject to paragraph (4), the new law governs all the issues specified in Article 2(1).

4. Except with respect to a person who has consented to a change of law, the old law continues to govern –

a) the existence of an interest in securities held with an intermediary arising before the change of law and the perfection of a disposition of those securities made before the change of law;

b) with respect to an interest in securities held with an intermediary arising before the change of law –

i) the legal nature and effects of such an interest against the relevant intermediary and any party to a disposition of those securities made before the change of law;

ii) the legal nature and effects of such an interest against a person who after the change of law attaches the securities;

iii) the determination of all the issues specified in Article 2(1) with respect to an insolvency administrator in an insolvency proceeding opened after the change of law;

c) priority as between parties whose interests arose before the change of law.

5. Paragraph (4)(c) does not preclude the application of the new law to the priority of an interest that arose under the old law but is perfected under the new law.

Article 8 Insolvency

1. Notwithstanding the opening of an insolvency proceeding, the law applicable under this Convention governs all the issues specified in Article 2(1) with respect to any event that has occurred before the opening of that insolvency proceeding.

2. Nothing in this Convention affects the application of any substantive or procedural insolvency rules, including any rules relating to –

a) the ranking of categories of claim or the avoidance of a disposition as a preference or a transfer in fraud of creditors; or

b) the enforcement of rights after the opening of an insolvency proceeding.

CHAPTER III – GENERAL PROVISIONS

Article 9 General applicability of the Convention

This Convention applies whether or not the applicable law is that of a Contracting State.

Article 10 Exclusion of choice of law rules (renvoi)

In this Convention, the term "law" means the law in force in a State other than its choice of law rules.

Article 11 Public policy and internationally mandatory rules

1. The application of the law determined under this Convention may be refused only if the effects of its application would be manifestly contrary to the public policy of the forum.

2. This Convention does not prevent the application of those provisions of the law of the forum which, irrespective of rules of conflict of laws, must be applied even to international situations.

3. This Article does not permit the application of provisions of the law of the forum imposing requirements with respect to perfection or relating to priorities between competing interests, unless the law of the forum is the applicable law under this Convention.

Article 12 *Determination of the applicable law for Multi-unit States*

1. If the account holder and the relevant intermediary have agreed on the law of a specified territorial unit of a Multi-unit State –

   a) the references to "State" in the first sentence of Article 4(1) are to that territorial unit;

   b) the references to "that State" in the second sentence of Article 4(1) are to the Multi-unit State itself.

2. In applying this Convention –

   a) the law in force in a territorial unit of a Multi-unit State includes both the law of that unit and, to the extent applicable in that unit, the law of the Multi-unit State itself;

   b) if the law in force in a territorial unit of a Multi-unit State designates the law of another territorial unit of that State to govern perfection by public filing, recording or registration, the law of that other territorial unit governs that issue.

3. A Multi-unit State may, at the time of signature, ratification, acceptance, approval or accession, make a declaration that if, under Article 5, the applicable law is that of the Multi-unit State or one of its territorial units, the internal choice of law rules in force in that Multi-unit State shall determine whether the substantive rules of law of that Multi-unit State or of a particular territorial unit of that Multi-unit State shall apply. A Multi-unit State that makes such a declaration shall communicate information concerning the content of those internal choice of law rules to the Permanent Bureau of the Hague Conference on Private International Law.

4. A Multi-unit State may, at any time, make a declaration that if, under Article 4, the applicable law is that of one of its territorial units, the law of that territorial unit applies only if the relevant intermediary has an office within that territorial unit which satisfies the condition specified in the second sentence of Article 4(1). Such a declaration shall have no effect on dispositions made before that declaration becomes effective.

Article 13 *Uniform interpretation*

In the interpretation of this Convention, regard shall be had to its international character and to the need to promote uniformity in its application.

Article 14 *Review of practical operation of the Convention*

The Secretary General of the Hague Conference on Private International Law shall at regular intervals convene a Special Commission to review the practical operation of this Convention and to consider whether any amendments to this Convention are desirable.

CHAPTER IV – TRANSITION PROVISIONS

Article 15 *Priority between pre-Convention and post-Convention interests*

In a Contracting State, the law applicable under this Convention determines whether a person's interest in securities held with an intermediary acquired after this Convention entered into force for that State extinguishes or has priority over another person's interest acquired before this Convention entered into force for that State.

Article 16 *Pre-Convention account agreements and securities accounts*

1. References in this Convention to an account agreement include an account agreement entered into before this Convention entered into force in accordance with Article 19(1). References in this Convention to a securities account include a securities account opened before this Convention entered into force in accordance with Article 19(1).

2. Unless an account agreement contains an express reference to this Convention, the courts of a Contracting State shall apply paragraphs (3) and (4) in applying Article 4(1) with respect to account agreements entered into before the entry into force of this Convention for that State in accordance with Article 19. A Contracting State may, at the time of signature, ratification, acceptance, approval or accession, make a declaration that its courts shall not apply those paragraphs with respect to account agreements entered into after the entry into force of this Convention in accordance with Article 19(1) but before the entry into force of this Convention for that State in accordance with Article 19(2). If the Contracting State is a Multi-unit State, it may make such a declaration with respect to any of its territorial units.

3. Any express terms of an account agreement which would have the effect, under the rules of the State whose law governs that agreement, that the law in force in a particular State, or a territorial unit of a particular Multi-unit State, applies to any of the issues specified in Article 2(1), shall have the effect that such law governs all the issues specified in Article 2(1), provided that the relevant intermediary had, at the time the agreement was entered into, an office in that State which satisfied the condition specified in the second sentence of Article 4(1). A Contracting State may, at the time of signature, ratification, acceptance, approval or accession, make a declaration that its courts shall not apply this paragraph with respect to an account agreement described in this paragraph in which the parties have expressly agreed that the securities account is maintained in a different State. If the Contracting State is a Multi-unit State, it may make such a declaration with respect to any of its territorial units.

4. If the parties to an account agreement, other than an agreement to which paragraph (3) applies, have agreed that the securities account is maintained in a particular State, or a territorial unit of a particular Multi-unit State, the law in force in that State or territorial unit is the law applicable to all the issues specified in Article 2(1), provided that the relevant intermediary had, at the time the agreement was entered into, an office in that State which satisfied the condition specified in the second sentence of

Article 4(1). Such an agreement may be express or implied from the terms of the contract considered as a whole or from the surrounding circumstances.

CHAPTER V – FINAL CLAUSES

Article 17 Signature, ratification, acceptance, approval or accession

1. This Convention shall be open for signature by all States.
2. This Convention is subject to ratification, acceptance or approval by the signatory States.
3. Any State which does not sign this Convention may accede to it at any time.
4. The instruments of ratification, acceptance, approval or accession shall be deposited with the Ministry of Foreign Affairs of the Kingdom of the Netherlands, Depositary of this Convention.

Article 18 Regional Economic Integration Organisations

1. A Regional Economic Integration Organisation which is constituted by sovereign States and has competence over certain matters governed by this Convention may similarly sign, accept, approve or accede to this Convention. The Regional Economic Integration Organisation shall in that case have the rights and obligations of a Contracting State, to the extent that that Organisation has competence over matters governed by this Convention. Where the number of Contracting States is relevant in this Convention, the Regional Economic Integration Organisation shall not count as a Contracting State in addition to its Member States which are Contracting States.
2. The Regional Economic Integration Organisation shall, at the time of signature, acceptance, approval or accession, notify the Depositary in writing specifying the matters governed by this Convention in respect of which competence has been transferred to that Organisation by its Member States. The Regional Economic Integration Organisation shall promptly notify the Depositary in writing of any changes to the distribution of competence specified in the notice in accordance with this paragraph and any new transfer of competence.
3. Any reference to a "Contracting State" or "Contracting States" in this Convention applies equally to a Regional Economic Integration Organisation where the context so requires.

Article 19 Entry into force

1. This Convention shall enter into force on the first day of the month following the expiration of three months after the deposit of the third instrument of ratification, acceptance, approval or accession referred to in Article 17.
2. Thereafter this Convention shall enter into force –
   a) for each State or Regional Economic Integration Organisation referred to in Article 18 subsequently ratifying, accepting, approving or acceding to it, on the

first day of the month following the expiration of three months after the deposit of its instrument of ratification, acceptance, approval or accession;

b) for a territorial unit to which this Convention has been extended in accordance with Article 20(1), on the first day of the month following the expiration of three months after the notification of the declaration referred to in that Article.

Article 20 Multi-unit States

1. A Multi-unit State may, at the time of signature, ratification, acceptance, approval or accession, make a declaration that this Convention shall extend to all its territorial units or only to one or more of them.
2. Any such declaration shall state expressly the territorial units to which this Convention applies.
3. If a State makes no declaration under paragraph (1), this Convention extends to all territorial units of that State.

Article 21 Reservations

No reservation to this Convention shall be permitted.

Article 22 Declarations

For the purposes of Articles 1(5), 12(3) and (4), 16(2) and (3) and 20 –

a) any declaration shall be notified in writing to the Depositary;

b) any Contracting State may modify a declaration by submitting a new declaration at any time;

c) any Contracting State may withdraw a declaration at any time;

d) any declaration made at the time of signature, ratification, acceptance, approval or accession shall take effect simultaneously with the entry into force of this Convention for the State concerned; any declaration made at a subsequent time and any new declaration shall take effect on the first day of the month following the expiration of three months after the date on which the Depositary made the notification in accordance with Article 24;

e) a withdrawal of a declaration shall take effect on the first day of the month following the expiration of six months after the date on which the Depositary made the notification in accordance with Article 24.

Article 23 Denunciation

1. A Contracting State may denounce this Convention by a notification in writing to the Depositary. The denunciation may be limited to certain territorial units of a Multi-unit State to which this Convention applies.
2. The denunciation shall take effect on the first day of the month following the expiration of twelve months after the date on which the notification is received by the

Depositary. Where a longer period for the denunciation to take effect is specified in the notification, the denunciation shall take effect upon the expiration of such longer period after the date on which the notification is received by the Depositary.

Article 24 Notifications by the Depositary

The Depositary shall notify the Members of the Hague Conference on Private International Law, and other States and Regional Economic Integration Organisations which have signed, ratified, accepted, approved or acceded in accordance with Articles 17 and 18, of the following –

a) the signatures and ratifications, acceptances, approvals and accessions referred to in Articles 17 and 18;

b) the date on which this Convention enters into force in accordance with Article 19;

c) the declarations and withdrawals of declarations referred to in Article 22;

d) the notifications referred to in Article 18(2);

e) the denunciations referred to in Article 23.

In witness whereof the undersigned, being duly authorised thereto, have signed this Convention.

Done at The Hague, on the ...... day of .... 20..., in the English and French languages, both texts being equally authentic, in a single copy which shall be deposited in the archives of the Government of the Kingdom of the Netherlands, and of which a certified copy shall be sent, through diplomatic channels, to each of the Member States of the Hague Conference on Private International Law as of the date of its Nineteenth Session and to each State which participated in that Session.

APPENDIX 2

CHRONOLOGY OF THE NEGOTIATIONS LEADING TO THE HAGUE SECURITIES CONVENTION

All the documents referred to are available on the website of the Hague Conference on Private International Law (under www.hcch.net)

- May 2000: Australia, the United Kingdom and the United States of America suggest that the HCCH develop, in a fast track procedure, a Convention on the law applicable to the dispositions of securities held with an intermediary; the Member States agree and recommend that the Conference start the preliminary work on such an instrument;
- November 2000: the Permanent Bureau issues a comprehensive Report, analysing the various issues to be addressed by the future Convention;[34]
- January 2001: an experts' meeting is held at The Hague to examine the possibility of preparing the Convention on a fast track procedure; the meeting is attended by 119 experts from 29 different Member States and 17 international organisations; the meeting produces a first tentative draft of key provisions of the future Convention;[35]
- May 2001: Drafting Committee meets in Paris
- June 2001: the HCCH holds the first part of its Nineteenth Diplomatic Session; among other things, the Members States officially endorse the project relating to securities held with an intermediary, including the fast-track procedure; it is decided to hold a Special Commission meeting in January 2002;
- July 2001: on the basis of comments received and the work of the Drafting Committee in Paris in May 2001, the Permanent Bureau circulates a new annotated draft on key provisions for a future Convention on this topic (the *annotated July 2001*

---

[34] C. BERNASCONI, *The Law Applicable to Dispositions of Securities Held Through Indirect Holding Systems,* Prel. Doc. No 1 of November 2000 for the attention of the Working Group of January 2001.

[35] For a summary of the results of this meeting, see *Report on the meeting of the Working Group of Experts (15 to 19 January 2001) and related informal work conducted by the Permanent Bureau on the law applicable to dispositions of securities held with an intermediary*, prepared by the Permanent Bureau, Prel. Doc. No 2 of June 2001 (this document was also presented as Prel. Doc. No 13 of June 2001 for the attention of the Nineteenth Session of the Hague Conference).

draft [36]); Member States and interested parties are invited to submit comments received on the draft before 1 October 2001;[37]

- October 2001: the Drafting Committee meets in Oxford to prepare a new intermediate draft of key provisions, which takes into account the numerous comments received on the annotated July 2001 draft; this draft (the *November 2001 draft* [38]) is sent for comments to all Member States, observers and any interested party;
- November 2001: the Permanent Bureau issues a Memorandum on federal clauses with new drafting proposals to be inserted in the future Convention;[39]
- November/December 2001: the Permanent Bureau holds a first series of 8 *Regional Discussion Workshops* (in Stockholm, Toronto, New York, Paris, Milan, Frankfurt, Hong Kong and Rio de Janeiro) to discuss and assess with State officials, legal experts and market participants the November 2001 draft;
- December 2001: the Drafting Committee meets in Brussels to reflect on the various comments made on the November 2001 draft and to prepare a new intermediate draft of key provisions for the future Convention;[40]
- January 2002: a *Special Commission* meets to finalise a new comprehensive draft Convention; this draft (the *January 2002 preliminary draft* [41]) is sent for comments to all Member States, observers and any interested party;[42]

---

36 *Tentative text on key provisions for a future Convention on the law applicable to proprietary rights in indirectly held securities, Suggestions for further amendment of the text contained in Working Document No 16 of the January 2001 experts' meeting*, submitted by the Permanent Bureau, Prel. Doc. No 3 of July 2001 for the attention of the Special Commission of January 2002.

37 See the *Chart summarising comments received on the "Annotated July 2001 draft"*, prepared by the Permanent Bureau, Prel. Doc. No 5 of November 2001 for the attention of the Special Commission of January 2002.

38 *Tentative text on key provisions for a future Convention on the law applicable to certain rights in respect of securities held with an intermediary, Suggestions for further amendment of the text contained in Preliminary Document No 3 of July 2001 (annotated July 2001 draft)*, submitted by the Permanent Bureau, Prel. Doc. No 6 of November 2001 for the attention of the Special Commission of January 2002.

39 *Memorandum on federal clauses with new drafting proposals for a future Hague Convention on the law applicable to certain rights in respect of securities held with an intermediary*, Prel. Doc. No 4 of November 2001, submitted by the Permanent Bureau.

40 *Tentative text on key provisions for a future Convention on the law applicable to certain rights in respect of securities held with an intermediary, Suggestions for further amendment of the text contained in Preliminary Document No 6 of November 2001 (November 2001 draft)*, submitted by the Permanent Bureau, Prel. Doc. No 7 of December 2001 for the attention of the Special Commission of January 2002.

41 *Preliminary draft Convention on the law applicable to certain rights in respect of securities held with an intermediary*, as adopted by the Special Commission on 17 January 2002 (provisional version), Prel. Doc. No 8 of February 2002, submitted by the Permanent Bureau.

42 *Chart summarising the comments received on the provisional version of the "preliminary draft Convention on the law applicable to certain rights in respect of securities held with an intermediary" as adopted by the Special Commission on 17 January 2002 (Prel. Doc. No 8)*, Prel. Doc. No 9 of March 2002, submitted by the Permanent Bureau.

- March 2002: the Drafting Committee meets in Frankfurt to reflect on the various comments made on the January 2002 preliminary draft and to prepare a new draft of key provisions (the *April 2002 preliminary draft* [43]); this draft is sent for comments to all Member States, observers and any interested party;[44]
- April 2002: the Permanent Bureau submits an Explanatory Note on the application of the future Convention with respect to Multi-unit States;[45]
- April 2002: The HCCH meets in Commission I on General Affairs and Policy of the Conference in the Nineteenth Diplomatic Session; the delegates unanimously commend the project for the progress achieved thus far and for the flexible working methods employed; it is decided to finalise the Convention on the basis of the working methods used thus far and, if possible, to hold a Diplomatic Conference before the end of 2002;[46]
- May 2002: the Permanent Bureau submits an Explanatory Note on the functioning of the Convention's conflict of laws rule with respect to transfers involving several intermediaries;[47]
- May 2002: the Permanent Bureau submits a proposal for a redraft of Articles 4 and 4 bis; it suggests that the Convention's primary conflict of laws rule should no longer focus on the concept of the place of the maintenance of an account, as this approach would not provide the required legal certainty and predictability; instead, the Permanent Bureau suggests to focus on the agreement between the relevant intermediary and the account holder and to submit this agreement to specific conditions;[48]
- May 2002: the Drafting Committee meets in London to reflect on the various comments made on the April 2002 preliminary draft and to prepare a new draft of key

---

43 *Preliminary draft Convention on the law applicable to certain rights in respect of securities held with an intermediary, Suggestions for amendment of the provisional version adopted by the Special Commission on 17 January 2002*, Prel. Doc. No 10 of April 2002, submitted by the Permanent Bureau.

44 See the *Chart summarising the comments received on the "April 2002 preliminary draft Convention on the law applicable to certain rights in respect of securities held with an intermediary" (Prel. Doc. No 10)*, Prel. Doc. No 14 of May 2002, submitted by the Permanent Bureau.

45 *Explanatory Note on Article 9 of the preliminary draft Convention on the law applicable to certain rights in respect of securities held with an intermediary, ("April 2002 preliminary draft" contained in Preliminary Document No 10)*, Prel. Doc. No 11 of April 2002, submitted by the Permanent Bureau.

46 See the *Conclusions of Commission I (General Affairs and Policy of the Conference) of the XIXth Diplomatic Session - April 2002.*

47 *Transfers involving several intermediaries: An Explanatory Note on the functioning of PRIMA within the framework of the preliminary draft Convention on the law applicable to certain rights in respect of securities held with an intermediary*, Prel. Doc. No 12 of May 2002, submitted by the Permanent Bureau. In response to this Note, the Japanese delegation to the Hague Securities project, submitted a document with *Comments on transfers involving two or more intermediaries: a response to Preliminary Document No 12*; this document was submitted as Prel. Doc. No 14A.

48 *Proposal for a redraft of articles 4 and 4 bis*, Prel. Doc. No 13 of May 2002, submitted by the Permanent Bureau.

provisions (the *June 2002 preliminary draft* [49]); this draft is sent for comments to all Member States, observers and any interested party;[50]

- June/July 2002: the Permanent Bureau holds a second series of 9 *Regional Discussion Workshops* (in Sydney, Tokyo, London, Copenhagen, Frankfurt, Rome, Paris, Toronto, New York) to discuss and assess the June 2002 preliminary draft with State officials, experts and markets participants;
- September 2002: the Permanent Bureau summarises the pros and cons of the various suggestions for the core provision of the future Convention (*i.e.*, the primary conflict of laws rule in Art. 4), reflecting the results of the discussions at the 9 Regional Discussion Workshops;[51]
- October 2002: the Permanent Bureau issues a Memo addressing the question of whether a specific provision is needed in the Convention to ensure that the applicable law is same for all dispositions of securities held with a particular intermediary, whether the disposition is in the form of a security interest, transfer of title by way of security or outright transfer;[52]
- November 2002: the Permanent Bureau issues an updated chart reflecting all the pros and cons of the various options suggested for the primary conflict of laws rule to be inserted in the Convention,[53] as well as a summary of the reactions to the Memo submitted in October 2002;[54]
- December 2002: the text of the draft Convention is adopted during a Diplomatic Session;
- February 2003: after some additional clean-up of the text adopted in December 2002, the Permanent Bureau releases the final text of the draft Convention.

---

49 *Preliminary draft Convention on the law applicable to certain rights in respect of securities held with an intermediary, Suggestions for amendment of the "April 2002 preliminary draft"*, Prel. Doc. No 15 of June 2002, submitted by the Permanent Bureau.

50 See the *Chart reflecting the comments received on the "Preliminary draft Convention on the law applicable to certain rights in respect of securities held with an intermediary" (Prel. Doc. No 15) and on "Options A and B in Article 4(1)" (Prel. Doc. No. 16),* Prel. Doc. No 18 of November 2002, submitted by the Permanent Bureau.

51 *Options A and B in Article 4(1): A brief summary of the results of the discussions at the 9 regional discussion workshops held in Asia, Europe and North America throughout June and July 2002*, Prel. Doc. No 16 of September 2002, submitted by the Permanent Bureau.

52 *Does the current draft of the Convention adequately ensure that the relevant intermediary [...] is the same for all dispositions of securities held with a particular intermediary, or is there a need for a specific provision to achieve this?,* Prel. Doc. No 17 of October 2002, submitted by the Permanent Bureau.

53 *Charts summarising the "pros and cons" of the Options A, A+ and B in Article 4(1),* Prel. Doc. No 18 bis of November 2002, submitted by the Permanent Bureau.

54 *Summary of reactions to Prel. Doc. No. 17: "Does the current draft of the Convention adequately ensure that the relevant intermediary [...] is the same for all dispositions of securities held with a particular intermediary, or is there a need for a specific provision to achieve this?"*, Prel. Doc. No 18 ter of November 2002, submitted by the Permanent Bureau.

APPENDIX 3

LIST OF ALL THE PRELIMINARY DOCUMENTS PRODUCED DURING THE NEGOTIATIONS LEADING TO THE HAGUE SECURITIES CONVENTION

(in reversed chronological order)

All the documents referred to are available on the website of the Hague Conference on Private International Law (under www.hcch.net). As to the nature and value of the *travaux préparatoires*, see the comment in paragraph Int-78.

- Preliminary Document No 18 ter of November 2002: Summary of reactions to Prel. Doc. No. 17: "Does the current draft of the Convention adequately ensure that the relevant intermediary [...] is the same for all dispositions of securities held with a particular intermediary, or is there a need for a specific provision to achieve this?"
- Preliminary Document No 18 bis of November 2002: Charts summarising the "pros and cons" of the Options A, A+ and B in Article 4(1)
- Preliminary Document No 18 of November 2002: Chart reflecting the comments received on the "Preliminary draft Convention on the law applicable to certain rights in respect of securities held with an intermediary" (Prel. Doc. No 15) and on "Options A and B in Article 4(1)" (Prel. Doc. No. 16)
- Preliminary Document No 17 of October 2002: Does the current draft of the Convention adequately ensure that the relevant intermediary [...] is the same for all dispositions of securities held with a particular intermediary, or is there a need for a specific provision to achieve this?
- Preliminary Document No 16 of September 2002: Options A and B in Article 4(1): A brief summary of the results of the discussions at the 9 Regional Discussion Workshops held in Asia, Europe and North America throughout June and July 2002
- Preliminary Document No 15 of June 2002: Preliminary draft Convention on the law applicable to certain rights in respect of securities held with an intermediary, Suggestions for amendment of the "April 2002 preliminary draft"
- Preliminary Document No 14A of May 2002: Comments on transfers involving two or more intermediaries: a response to Preliminary Document No 12, submitted by the Japanese delegation
- Preliminary Document No 14 of May 2002: Chart summarising the comments received on the "April 2002 preliminary draft Convention on the law applicable to certain rights in respect of securities held with an intermediary" (Prel. Doc. No 10)
- Preliminary Document No 13 of May 2002: Proposal for a redraft of articles 4 and 4 bis

- Preliminary Document No 12 of May 2002: Transfers involving several intermediaries: An Explanatory Note on the functioning of PRIMA within the framework of the preliminary draft Convention on the law applicable to certain rights in respect of securities held with an intermediary
- Preliminary Document No 11 of April 2002: Explanatory Note on article 9 of the preliminary draft Convention on the law applicable to certain rights in respect of securities held with an intermediary, ("April 2002 preliminary draft" contained in Preliminary Document No 10)
- Preliminary Document No 10 of April 2002: Preliminary draft Convention on the law applicable to certain rights in respect of securities held with an intermediary - Suggestions for amendment of the provisional version adopted by the Special Commission on 17 January 2002
- Preliminary Document No 9 of March 2002: Chart summarising the comments received on the provisional version of the "preliminary draft Convention on the law applicable to certain rights in respect of securities held with an intermediary" as adopted by the Special Commission on 17 January 2002 (Prel. Doc. No 8)
- Preliminary Document No 8 of February 2002: Preliminary draft Convention on the law applicable to certain rights in respect of securities held with an intermediary, as adopted by the Special Commission on 17 January 2002
- Preliminary Document No 7 of December 2001: Tentative text on key provisions for a future Convention on the law applicable to certain rights in respect of securities held with an intermediary. Suggestions for further amendment of the text contained in Preliminary Document No 6
- Preliminary Document No 6 of November 2001: Tentative text on key provisions for a future Convention on the law applicable to certain rights in respect of securities held with an intermediary
- Preliminary Document No 5 of November 2001: Chart summarising comments received on the "Annotated July 2001 draft"
- Preliminary Document No 4 of November 2001: Memorandum on federal clauses with new drafting proposals for a future Hague Convention on the law applicable to certain rights in respect of securities held with an intermediary
- Preliminary Document No 3 of July 2001: Tentative text on key provisions for a future Convention on the law applicable to proprietary rights in indirectly held securities - Suggestions for further amendment of the text contained in Working Document No 16 of the January 2001 experts meeting
- Preliminary Document No 2 of June 2001: Report on the meeting of the Working Group of Experts (15 to 19 January 2001) and related informal work conducted by the Permanent Bureau on the law applicable to dispositions of securities held with an intermediary
- Preliminary Document No 1 of November 2000: Report on the Law Applicable to Dispositions of Securities Held Through Indirect Holding Systems, prepared by Christophe Bernasconi

APPENDIX 4

LIST OF ALL EXPERTS WHO PARTICIPATED IN THE DIPLOMATIC SESSION HELD FROM 2-13 DECEMBER 2002

PRESIDENT OF THE NINETEENTH SESSION OF THE HAGUE CONFERENCE ON PRIVATE INTERNATIONAL LAW

Mr *Antoon (Teun) V.M. Struycken* (Netherlands)

CHAIR OF COMMISSION III (mandated to prepare the Securities Convention)

Mrs *Stefania Bariatti* (Italy)

REPORTERS

Mr *Roy Goode* * (United Kingdom of Great Britain and Northern Ireland)
Mr *Hideki Kanda* * (Japan)
Mr *Karl Kreuzer* * (Germany)

DRAFTING COMMITTEE

The Members of the Drafting Committee who served during the Diplomatic Session are identified by an asterisk (*). The Drafting Committee was chaired by Professor Sir Roy Goode. See also footnote 9 of the Report.

DELEGATES

Argentina
Mr *Rubén Vallejo*, Counsellor, Embassy of Argentina, The Hague
Ms *Maria del Carmen Urquiza*, Head of Legal Department, Central Bank of Argentina, Buenos Aires
Mr *Luis Wetzler*, Legal Counsel in International Financial Affairs, Central Bank of Argentina, Legal Department, Buenos Aires
Ms *Inés Weinberg de Roca*, Judge, Professor of private international law, University of Buenos Aires, Buenos Aires

Australia
Mr *Matt Minogue*, Assistant Secretary, Civil Justice Division, Attorney-General's Department, Barton
Mr *Ian Beckett*, Financial System Division, Department of the Treasury, Canberra
Mr *Richard Potok*,* Legal Adviser to the Permanent Bureau; Potok & Co, Sydney

Austria
Mr *Martin Adensamer*, Director, Federal Ministry of Justice, Vienna
Mr *Peter Pöch*, Legal Counsellor, Austrian Central Securities Depository, Vienna

Belgium
Mr *Jean-Pierre Deguée*, Conseiller, Commission Bancaire et Financière, Bruxelles
Mr *Diego Devos*,* Director, Deputy General Counsel, Euroclear Bank S.A./N.V., Bruxelles

Brazil
H.E. Mr *Affonso de Alencastro Massot*, Ambassador of Brazil, Embassy of Brazil, The Hague
Ms *Janine-Monique Bustani*, Counsellor, Embassy of Brazil, The Hague
Mr *Fábio Abud Antibas*, Secretary, Embassy of Brazil, The Hague
Mr *João Lauro Amaral*, Managing Director, Brazilian Mercantile & Futures Exchange (USA) Inc., New York

Bulgaria
Ms *Tzvety Romanska*, Attaché, Ministry of Foreign Affairs, International Law Department, Sofia

Canada
Ms *Manon Dostie*, Counsel, Private International Law Team, Public Law Policy Section, Department of Justice, Ottawa
Bradley Crawford, Q.C., McCarthy Tétrault, Toronto
Mr *Michel A. Brunet*, Lawyer, Fraser Milner Casgrain, Montréal
Mr *Eric Spink*, Vice Chair, Alberta Securities Commission, Edmonton
Mr *Gregg A. Calkin*, Counsellor, Embassy of Canada, The Hague

Chile
Mr *Alejandro Sfeir-Tonsic*, Second Secretary, Embassy of Chile, The Hague

China
H.E. Mr *Zhu Zushou*, Ambassador of China, Embassy of the People's Republic of China, The Hague
Mrs *Zhang Honghong*, Counsellor, Department of Treaty and Law, Ministry of Foreign Affairs, Beijing
Ms *Lulu Zhou*, Attaché, Department of Treaty and Law, Ministry of Foreign Affairs, Beijing
Mr *Yu Fan*, Legal Official, Department of Legal Affairs, China Securities Regulatory Commission, Beijing
Mr *Sun Guoshun*, First Secretary, Embassy of the People's Republic of China, The Hague
Mr *James Ding*, Government Counsel, International Law Division, Treaties & Law Unit, Department of Justice of Hong Kong Special Administrative Region, Hong Kong

Czech Republic
Mr *Jan Čížek*, Counsellor, Deputy Head of Mission, Embassy of the Czech Republic, The Hague
Mr *Jiří Špička*, Director of the Financial Market and Banking Department, Ministry of Finance of the Czech Republic, Prague
Mrs *Věra Krajánková*, Head of the Department of International Legal Aid in Civil Matters, Ministry of Justice, Prague

Denmark
Mr *Thorkild Fodge*, Head of Division, Law Department, Ministry of Justice, Copenhagen
Mr *Kristian Korfits Nielsen*, Head of Section, Law Department, Ministry of Justice, Copenhagen
Mr *Ulrik Rammeskow Bang-Pedersen*,* Professor of Law, University of Copenhagen, Faculty of Law, Copenhagen

Egypt
Ms *Namira Negm*, Second Secretary, Embassy of the Arab Republic of Egypt, The Hague

Finland
Mrs *Marja A. Tuokila*, Legal Adviser, Law Drafting Department, Ministry of Justice, Helsinki
Mr *Janne Lauha*, Senior Vice President, HEX Group, Legal Unit, Helsinki

France
Ms *Michelle Dubrocard*, Conseiller juridique, Ambassade de France, La Haye
Mr *Michel Germain*, Professor, Université de Paris II Panthéon-Assas, Vincennes
Mr *Stéphane Beraud*, Adjoint de Direction, Direction des services juridiques, Banque de France, Paris
Mr *Alexandre de Fontmichel*,* Lawyer, Ministry of Justice, Paris

Germany
Ms *Ute Hoehfeld*, Head of Section for Law relating to Securities, Banks and the Stock Exchange, Federal Ministry of Justice, Berlin
Mr *Fabian Reuschle*,* Richter, Executive Assistant, Section on Law relating to Co-operatives, Banks and the Stock Exchange; Law relating to Securities, Federal Ministry of Justice, Berlin
Mr *Olaf Christmann*, Bundesbankrat, Legal Department, Deutsche Bundesbank, Frankfurt am Main
Mr *Jürgen Becker*, Third Secretary, Legal & Social Affairs Section, Embassy of Germany, The Hague

Ireland
Mr *Christopher O'Toole*, Advisory Counsel, Office of the Attorney General, Dublin

Italy
Mr *Gianfranco Trovatore*, Legal Counsel, CONSOB, Markets and Economic Analysis Department, Rome

Japan
Mr *Yoshihisa Hayakawa*, Associate Professor, Faculty of Law, Rikkyo University, Tokyo
Mr *Masami Hadama*, Attorney, Civil Affairs Bureau, Ministry of Justice, Tokyo
Mr *Hironori Wanami*, Attorney, Civil Affairs Bureau, Ministry of Justice and Treaties Bureau, Ministry of Foreign Affairs, Tokyo
Mr *Kunio Koide*, First Secretary, Embassy of Japan to the Netherlands, The Hague
Mr *Taisaku Ikeshima*, Legal Researcher / Adviser, Embassy of Japan to the Netherlands, The Hague

Republic of Korea
Mr *Chong-Hoon Kim*, Counsellor, Embassy of Korea, The Hague
Mr *Ki-Jin You*, Second Secretary, Embassy of the Republic of Korea to the Italian Republic, Rome

Luxembourg
Mr *Daniel Ruppert*, Attaché de Gouvernement 1er en rang auprès du Ministère de la Justice, Luxembourg
Mr *Philippe Dupont*,* Avocat à la Cour, Arendt & Medernach, Luxembourg

Malaysia
Ms *Lee Foong Mee*, Senior Manager, Legal Department, Bank Negara Malaysia (Central Bank of Malaysia), Kuala Lumpur
Ms *Benothini Bascaran*, Manager, Law Review & Development, Corporate Affairs, Legal & Compliance, Clearing, Settlement & Depository Group, Kuala Lumpur
Mrs *Sujatha Sekhar Naik*, Senior Executive, Law Reform & Regulatory Policy Department, Securities Commission, Kuala Lumpur

Mexico
Mrs *Zadalinda González y Reynero*, Minister, Embassy of Mexico, The Hague
Mr *Leonel Pereznieto Castro*, Adviser to the Legal Office of the Ministry of Foreign Affairs of Mexico, Official Counsel of Von Wobeser y Sierra, SC, Mexico

Netherlands
Mr *Antoon (Teun) V.M. Struycken*, président de la Commission d'Etat néerlandaise de droit international privé, Professeur de droit émérite, Heilig Landstichting
Mrs *Dorothea van Iterson*, Counsellor of Legislation Department of Private Law, Ministry of Justice, The Hague
Mr *Willem (Pim) Rank*, Attorney at Law, Nauta Dutilh, Amsterdam, Professor of Securities Law, University of Nijmegen, Nijmegen

Panama
Mr *Pedro Sitton*, Legal Counsellor, Embassy of Panama, Brussels

Portugal
H.E. Mr *João Salgueiro*, Ambassador of Portugal, Embassy of Portugal, The Hague
Ms *Raquel Correia*, Legal Adviser, Ministry of Justice, Lisbon

Mrs *Rafaela Rocha*, Legal Adviser, Comissão do Mercado de Valores Mobiliários (Securities Regulator), Lisbon

Russian Federation
H.E. Mr *Alexandre Khodakov*, Ambassador of the Russian Federation, Embassy of the Russian Federation, The Hague
Mr *Gennady Kuzmin*, First Secretary, Embassy of the Russian Federation, The Hague
Mr *Igor Panin*, First Secretary, Embassy of the Russian Federation, The Hague
Mr *Kirill Sokolov-Shcherbachev*, Attaché, Embassy of the Russian Federation, The Hague

Spain
Mr *Francisco J. Garcimartín Alférez*,* Professor of Private International Law, Universidad de Castilla-La Mancha, Faculty of Law, Madrid
Mrs *Maria González Fernandez*, Técnico Superior, General Directorate for Legislative Policy and International Judicial Cooperation, Ministry of Justice, Madrid

Sweden
Mr *Lars Afrell*,* Director, Financial Institution and Market, Ministry of Finance, Stockholm
Ms *Amina Lundqvist*, Legal Adviser, Ministry of Justice, Stockholm

Switzerland
Mr *Alexander Markus*, Head of Section on Private International Law, Federal Office of Justice, Bern
Mr *Samuel Baumgartner*, Deputy Head, Section on Private International Law, Federal Office of Justice, Bern
Mr *Daniel Girsberger*, Professor, University of Lucerne, Lucerne
Mr *Hans Kuhn*, Head of Legal Service, Swiss National Bank, Zurich
Mr *Nedim Peter Vogt*, Attorney at Law, Bär & Karrer, Zurich

United Kingdom of Great Britain and Northern Ireland
Mr *Guy Morton*,* Freshfields Bruckhaus Deringer, London
Mr *James Neilson*, Legal Adviser, Her Majesty's Treasury, London

United States of America
Mr *Harold S. Burman*, Executive Director, Advisory Committee on Private International Law, Office of the Legal Adviser, Department of State, Washington DC
Ms *Joyce Hansen*,* Deputy General Counsel and Senior Vice President, Federal Reserve Bank of New York, New York
Ms *Cynthia Reese*, Deputy Chief Counsel, Bureau of the Public Debt, Department of Treasury, Washington DC
Mr *Jack R. Wiener*, Managing Director and Deputy General Counsel, The Depository Trust & Clearing Corporation, New York
Mr *Harry C. Sigman*, Commercial Law Adviser, Los Angeles
Mr *James S. Rogers*,* Professor, Boston College Law School, Newton

OBSERVERS

REPRESENTATIVES FOR INVITED STATES

Lebanon
Ms *Rana Mokkaddem*, First Secretary, Embassy of Lebanon, The Hague

Ukraine
Ms *Lidia Horbunova*, Deputy State Secretary, Ministry of Justice, Kiev
Mr *Anatoly Dovgert*, Professor, Head of the Department of Private International Law, Kiev National University, Kiev
Mr *German Galuschenko*, Chief Legal Adviser, Presidential Office, Kiev
Ms *Kateryna Shevchenko*, Director of the Department of International Legal Cooperation and Legal Assistance, Ministry of Justice, Kiev
Mr *Vsevolod Chentsov*, Head of Treaty Department, Legal Directorate General, Ministry of Foreign Affairs, Kiev

REPRESENTATIVES FOR INTERGOVERNMENTAL ORGANISATIONS

European Central Bank
Mr *Antonio Sáinz de Vicuña*, General Counsel, Directorate General Legal Services, Frankfurt am Main
Mr *Klaus Löber*, Senior Legal Counsel, Frankfurt am Main

European Investment Bank
Mr *Marc Dufresne*, Director, Legal Affairs Directorate, Luxembourg
Ms *Nicola Barr*, Head of Division, Legal Affairs Directorate, Luxembourg
Mr *Pierre Albouze*, Head of Division, Legal Affairs Directorate, Luxembourg

United Nations Commission on International Trade Law
Mr *Spiros V. Bazinas*, Senior Legal Officer, UNCITRAL Secretariat, International Trade Law Branch, Office of Legal Affairs, Vienna

European Commission
Mr *José María Fombellida*, Head of Unit, Financial instruments, European Commission, Brussels
Mr *Jean-Luc Filippini*, Administrator, European Commission, Internal Market DG, Brussels
Mr *Jens Schröder*, Administrator, European Commission, Internal Market DG, Brussels
Mr *Jérôme Carriat*, Administrator, European Commission, Brussels

Council of the European Union
Mr *Fernando Paulino Pereira*, Administrateur principal à la Direction générale "H", Secrétariat général, Bruxelles
Mrs *Nathalie Pensaert*, Administrator, Directorate General H, Justice & Home Affairs, General Secretariat, Brussels
Ms *Anne Fisker*, Head of Section, Law Department, Ministry of Justice, Copenhagen

International Institute for the Unification of Private Law
Mr *J. Herbert E. Kronke*, Secretary General, Rome
Mr *Philipp Paech*, Research Officer, Rome

International Organization of Securities Commissions
Ms *Andrea M. Corcoran*, Director, Office of International Affairs, U.S. Commodity Futures Trading Commission, Chair, Implementation Committee, IOSCO, Washington DC

European Parliament
Mr *Harry Duintjer Tebbens*, Head of Division, Legal Service, Brussels

REPRESENTATIVES FOR NON-GOVERNMENTAL ORGANISATIONS

American Central Securities Depositories Association
Ms *Margarida A. C. Baptista*, Working Committee Member, Economist, CBLC Brazilian Clearing & Depository Corporation, Sao Paulo

Emerging Markets Traders Association
Ms *Sandra M. Rocks*, Cleary, Gottlieb, Steen & Hamilton, New York

European Financial Market Lawyers Group
Mr *Helmut Merkel*, Dresdner Bank AG, Frankfurt am Main

European Banking Federation
Mr *Philippe Langlet*, Head of Global Securities Services, Société Générale, Paris
Ms *Annette Maas*, Adviser, Financial Markets, Bundesverband deutscher Banken, Berlin

Financial Market Lawyers Group
Mr *Locke McMurray*, Director & Senior Counsel, Global Markets & Investment Banking, Merrill Lynch, Office of General Counsel, New York
Mr *Robert Sussman*, Managing Counsel, The Bank of New York, Legal Department, New York

International Bar Association
Ms *Lisa Curran*, Of Counsel, Allen & Overy, Rome
Mr *Randall D. Guynn*,* Partner, Davis Polk & Wardwell, New York
Mr *Antoine Maffei*,* avocat à la Cour, De Pardieu Brocas Maffei & Leygonie, Paris
Mr *Akihiro Wani*, Attorney at Law, Partner, Mitsui Yasuda Wani & Maeda, Tokyo

International Committee of the Uniform Law Commissioners
Mr *Curtis R. Reitz*, Chair of the Committee on International Legal Developments of the National Conference of Commissioners on Uniform State Laws, Chicago

International Councils of Securities Associations
Mr *Clifford Dammers*, Secretary General, International Primary Market Association, London
Ms *Nancy Marchand*, Avocat à Cour, Counsel, Davis Polk & Wardwell, Paris

International Swaps and Derivative Association
Mr *Lawrence Brandman*, Associate General Counsel, Goldman Sachs & Co., New York
Mr *Karl H. Kircher*, Senior Counsel, Deutsche Bank, Frankfurt am Main
Mr *Peter Martin Werner*, Assistant Director of European Policy, International Swaps & Derivatives Association, Inc. London
Mr *David Bloom*, Head of Legal, Corporate, Investment Banking, HSBC Holdings plc., London

SECRETARIAT

Secretary General
Mr *Hans van Loon*

Deputy Secretary General
Mr *William Duncan*

First Secretaries
Mr *Christophe Bernasconi* *
Mr *Philippe Lortie*
Ms *Andrea Schulz*

Recording Secretaries
Ms *Eleanor Cashin-Ritaine*
Ms *Florence Guillaume*
Ms *Kara Iland*, Lawyer
Mr *Anatoly Kirievsky*
Mr *Kostya Rilov*
Ms *Laurence Thébault*

INDEX BY SUBJECT

Figures refer to paragraph numbers.

A

Acceptance of the Convention, 17-1 *et seq.*, 18-1 *et seq.*

Accession to the Convention, 17-1 *et seq.*, 18-1 *et seq.*

Account agreement (relevant)

Basis for determining applicable law, *See Primary rule*
Definition, 1-15
Disposition in favour of the intermediary, 4-41, 16-8
Amendment, Int-65, 4-27 *et seq.*, 7-1 *et seq.*
New account agreement, 7-11
Pre-Convention account agreement, Int-75, 16-1 *et seq.*

Account holder

Definition, 1-13, 1-14
Rights
- Nature of rights, Int-22, Int-23, 2-1 *et seq.*, 2-11 *et seq.*
- Scope of the Convention, Int-24, Int-25, 2-11 *et seq.*

Accounting requirement (irrelevant in determining location of securities account), Int-43, 4-26

Adverse claim, *See also Competing interest*

Assertion of, 2-26

Amendment of the account agreement, *See Acount agreement, Change of applicable law (Art. 7)*

American Depositary Receipt (ADR), 1-2, and *See Securities (definition)*

Applicable law

Change of applicable law, *See this expression*
Determination of the applicable law under the Convention, *See Primary rule, Fall-back rules*
Exception (non-application), *See Public policy, Internationally mandatory rules*
Express choice of law, 4-17, 4-18, 16-20
Factors to be disregarded, 6-1 *et seq.*
Insolvency proceeding, *See this expression*
Law in force, 4-15, 12-12, 12-13
Non-Contracting State, 9-1
Qualifying Office requirement, *See this expression*
Renvoi (exclusion), 10-1 *et seq.*
Scope, *See Article 2(1) issues*

Approval of the Convention, 17-1 *et seq.*, 18-1 *et seq.*

Article 2(1) issues, Int-54 to Int-59, 1-42, 2-1 *et seq.*

Applicable law (determination), *See Primary rule, Fall-back rules, Change of applicable law*
Competing interest (intermediary's duties), 2-24 *et seq.*
Contractual rights, Int-55, 2-2, 2-4 *et seq.*, 2-11, 2-30 *et seq.*
Dispositions (nature and effect), 2-18 *et seq.*

Dividends, income, redemption, sale proceeds etc. (entitlement to), 2-29
Nature of interest in securities, 2-4, 2-5, 2-7, 2-8, 2-11 *et seq.*
Nature of the list of issues, 2-9 *et seq.*
Perfection requirements, 2-21, 2-22, *See also Perfection*
Priority, 2-23, *See also Priority*
Purely contractual or otherwise purely personal rights (exclusion), Int-59, 2-7, 2-17, 2-32 *et seq.*
Realisation of interest in securities, 2-28
Right resulting from a credit of securities (nature and effect), 2-11 *et seq.*
Rights and duties of an issuer of securities, registrar or transfer agent (exclusion), 2-34 *et seq.*
Upper-tier attachment, *See this expression*

B

Bankers acceptances, 1-2 and *See Securities (definition)*

Bearer securities, Int-36, 1-1

Bonds (listed or not), 1-2 and *See Securities (definition)*

Book entry, Int-29 *et seq.*

C

Cash (exclusion), *See Securities*

Central Securities Depository (CSD)

Declaration (Art. 1(5)), 1-37 *et seq.*, 22-1 *et seq.*
Intermediary, Int-17 *et seq.*, 1-32, 1-36, 1-37 *et seq.*
International (ICSD), Int-17 *et seq.*
Participants (of CSD or ICSD), Int-18

Centralisation of securities, Int-21

Certificates of deposit, 1-2 and *See Securities (definition)*

Chain of intermediaries (transfer through), Int-63, 4-43 *et seq.*

Change of applicable law (Art. 7), Int-65, 4-27 *et seq.*, 7-1 *et seq.*

Conditions and scope, 7-8 *et seq.*
Effect, 7-15 *et seq.*
- Protected person, 7-16 *et seq.*
- Person who attaches, 7-21
- Insolvency administrator, 7-21

Exceptions to general rule, 7-16 *et seq.*
General rule (see also exceptions), 7-15
Priority issues, 7-22 *et seq.*, 7-25 *et seq.*

Characterisation, *See also Recharacterisation of disposition*

Account holder rights's, Int-22, Int-23, 2-1 *et seq.*, 2-5, 2-11 *et seq.*

CHESS (Australia), *See Intermediary*

Choice of governing law, *See Primary rule, Change of the applicable law*

Collateral, *See Disposition (definition)*

Competing interest

Duties of the intermediary (applicable law), 2-24 *et seq.*
Priority, *See this expression*
Upper-tier attachment (applicable law), 2-26

Conflict of laws rule

Fall-back rules, *See this expression*
General applicability (universal character), 9-1
Internationality, *See this expression*

Lex rei sitae principle (criticism), Int-36 *et seq.*, 4-3 *et seq.*
Look-through approach (rejection), Int-38 *et seq.*, 4-5
Need to reflect market reality, Int-2 *et seq.*, Int-33 *et seq.*
Place of the Relevant Intermediary Approach (PRIMA) (content and abandonment), Int-41 *et seq.*, 4-4 *et seq.*
Primary rule, *See this expression*
Renvoi, *See this expression*
Super-PRIMA (rejection), Int-63, 4-5, 4-11, 4-43 *et seq.*

Contracting State, 1-43, 17-1 *et seq.*

Contractual right (applicable law), Int-55, 2-2, 2-4 *et seq.*, 2-11, 2-30 *et seq.*, *See also Purely contractual or otherwise purely personal right*

Convention law, *See Applicable law*

Corporate actions (meaning), 4-36

Credit of securities (to an account)

Nature and effects (applicable law), 2-11 *et seq.*
Book entry, *See this expression*

CREST (United Kingdom), *See Intermediary*

D

Declaration, 22-1 *et seq.*

Multi-unit State,
Qualifying office, 12-21 *et seq.*
Fall-back rules, Int-72, 12-16 *et seq.*
Territorial scope, 20-1 *et seq.*
Pre-Convention account agreement, 16-13, 16-16 *et seq.*
Modification, 22-3
Operator of a system (Art. 1(5)), 1-37 *et seq.*
Time effect, 22-5
Withdrawal, 22-4

Dematerialisation, Int-21

Denunciation of Convention, 23-1 *et seq.*, 24-1 *et seq.*

Depositary of the Convention (Notifications by), 24-1 *et seq.*

Derivatives, 1-2 and *See Securities (definition)*

Diplomatic Session (Nineteenth), Int-15

List of Participants, *Appendix 4*

Direct holding

French law (treated as direct holding falling within scope of the Convention), 1-16
Japanese law (treated as direct holding falling within scope of the Convention), 1-16
Move from direct holdings to intermediated holdings, Int-17 *et seq.*

Disposition (of securities or securities account)

Definition, 1-19, 1-30 *et seq.*
Disposition by account holder to the intermediary itself, 4-41 *et seq.*
Disposition by transfer from one securities account to another, incl. through chain of intermediaries, Int-63, 4-43 *et seq.*
Legal nature and effects (applicable law), 2-18 *et seq.*
Lien by operation of law, 1-31, *See also this expression*
Outright transfer of title, Int-36, 1-19, 1-30, 2-18, 2-20

Perfection requirements (applicable law), 2-21 *et seq.*, 7-16 *et seq.*, 7-21 *et seq.*, 11-12
Pre-Convention dispositions (transitions provisions), 15-1 *et seq.*
Repo, 1-19, 2-8, 2-19, 11-7
Transfer of title by way of security, Int-36, 1-19, 2-18 *et seq.*, 11-7

Distribution (entitlement to) (applicable law), 2-29

Dividends (entitlement to) (applicable law), 2-29

E

Entry into force (of the Convention), 19-1 *et seq.*

Depositary of the Convention (notifications by), *See this expression*
Gap period, Int-76, 16-3, 16-11 *et seq.*, 16-13, 16-22
International plan (on the), 19-1 *et seq.*, 19-4 *et seq.*
State or REIO, 19-1 *et seq.*, 16-6 *et seq.*

F

Factors to be disregarded when determining the applicable law, 3-3, 6-1 *et seq.*

Fall back rules (Art. 5), Int-62, 5-1 *et seq.*

Amendment of the account agreement, 5-12, 4-27 *et seq.*, 7-1 *et seq.*
Express and unambiguous choice, 5-4 *et seq.*
First fall-back rule, 5-4 *et seq.*
Multi-unit State, 5-3, 12-16 *et seq.* (declaration preservering internal conflict of laws rules), *See also Multi-unit State*
Operation as a cascade, 4-9, 5-1
Second fall-back rule, 5-7 *et seq.*
Third fall-back rule, 5-11
Time factor, 5-12, 7-13
Written account agreement, 5-4 *et seq.*

Financial futures, 1-2 and *See Securities (definition)*

Financial instruments, *See Securities (definition)*

Financial options, 1-2 and *See Securities (definition)*

Fraud of law, 3-10, 11-5

Fungible account, *See also Pooling of Securities*

Rights in, Int-39, 2-11

G

Gap period, *See Entry into force (of the Convention)*

General applicability of Convention, 9-1

Global (jumbo) certificate, 1-2

Global note, *See Global (jumbo) certificate*

Governing law, *See Applicable law*

Governing law provision, *See Primary rule, Change of applicable law*

Gutschrift in Wertpapierrechnung, Int-23, 2-11, *See Account holder (rights, nature of the right)*

H

History of the Convention, Int-1 *et seq.*

Adoption
  Negotiation process, Int-6 *et seq.*
  Fast-track procedure, Int-6 *et seq.*
  Consensus, Int-9, Int-14
  Transparency, Int-11
  Travaux préparatoires (nature of), Int-78
Goals and purpose, Int-16 *et seq.*, Int-50

Holding of securities

Direct holding, Int-16, Int-17
Intermediated holding, Int-16, Int-18 *et seq.*, *See also Scope of the Convention*

I

Immobilisation of security certificates, Int-21

Income (entitlement to) (applicable law), 2-29

Insolvency, *See Insolvency proceeding*

Insolvency administrator

Definition, 1-27
Old law (application in relation to), 7-21

Insolvency proceeding

Acquired interests (effects of) in insolvency proceeding, Int-66, 8-9 *et seq.*
Acquired interests (recognition of) in insolvency proceeding, Int-66, 8-7 *et seq.*
Applicable law, Int-66, 8-1 *et seq.*, 11-4
  Change of applicable law (effect), 7-21, *See also this expression*
Avoidance provisions not affected, 8-4
Definition, 1-26
Effect of, 8-1 *et seq.*
Insolvency administrator, *See this expression*
Lex concursus, Int-66
Must respect Convention rules
  Acquisition (on), 8-7, 8-9
  Perfection (on), 8-4 *et seq.*
  Priorities (on), 8-4
Ranking of claims not affected, 8-6

Interest in Securities (meaning of), 1-4

Intermediary, *See also Relevant intermediary*

Account manager, agent or other, 1-35
Central bank, 1-36
CHESS System, 1-34, 1-36
CREST System, 1-32, 1-34, 1-36, 1-37 *et seq.*
CSD, ICSD, *See Central Securities Depositary (CSD)*
Declaration (Art. 1(5)), 1-37 *et seq.*
Definition, 1-10 *et seq.*, 1-32 *et seq.*
Duties of an intermediary against person who asserts competing interest (incl. *upper-tier attachments*), 2-24 *et seq.*
Nordic system, 1-32 *et seq.*, 1-36
Operator of a system, *See this expression*
Registrar agent, 1-33 *et seq.*
Transfer agent, 1-33 *et seq.*

Intermediated holding pattern

Move from direct holdings to intermediated holdings, Int-17 *et seq.*
Typical pattern (examples), Int-26, Int-27, Int-31

Internal renvoi, *See Multi-unit State*

International Central Securities Depositary (ICSD), *See Central Securities Depositary*

Internationality, Int-53, 3-1 *et seq.*

Factors to be disregarded, 3-3, 6-1 *et seq.*
Fraud of law, 3-10
Illustrations, 3-6 *et seq.*
Multi-unit State, 3-13 *et seq.*
Time, 3-11 *et seq.*

Internationally mandatory rules (application), Int-68, 11-1 *et seq.*, 11-9 *et seq.*, 11-12 (limitation)

Interpretation of the Convention (uniform and autonomous (Art. 13)), Int-73, 13-1 *et seq.*

Islamic, Securities issued under Islamic law, 1-2 and *See Securities (definition)*

Issuer (rights and duties of), 2-34

K

Key features of the Convention, Int-49 *et seq.*

L

Lex concursus, See Insolvency proceeding

Lex rei sitae, See Conflict of laws rule

Lien by operation of law, 1-31, 11-12

Look-through approach, *See Conflict of laws rule*

M

Mandatory rules, *See Internationally mandatory rules*

Monitoring of the practical operation of Convention, 14-1 *et seq.*

Multi-unit State,

Conflict of laws between territorial units, 3-13 *et seq.*
Declaration, Int-72, 12-16 *et seq.*, 12-21 *et seq.*, 20-1 *et seq.*
Definition, Int-69, 1-28, 12-1
Fall-back rules, 12-16 *et seq.*
Internal conflict of laws rules (Internal Renvoi), Int-71 *et seq.*, 10-2, 12-14 *et seq.*, 12-16 *et seq.*
Law in force in a territorial unit, Int-70, 4-15, 10-4, 12-12 *et seq.*
Primary rule, 12-6 *et seq.*, 12-14, *See also this expression*
Qualifying Office requirement, 12-10 *et seq.*, 12-21 *et seq.*
Territorial scope, 20-1 *et seq.*

N

'New law' (in case account agreement is amended so as to change applicable law under the Convention)

Application of, 7-15, 7-17, 7-20
Definition, 7-14

O

Office

Definition, 1-22 *et seq.*
Place of business, 1-23 *et seq.*
Qualifying Office requirement, *See this expression*
Temporary place of business, 1-24

'Old law' (in case account agreement is amended so as to change applicable law under the Convention)

Application of, 7-16 *et seq.*
Definition, 7-14

Omnibus account, *See Fungible account, Pooling of Securities*

Operator of a system, 1-37 *et seq.*

P

Perfection (of a disposition)

Applicable law (requirements), 2-21 *et seq.*
Change of the applicable law, 7-19 *et seq.*
Definition, 1-20 *et seq.*
Insolvency proceedings, *See this expression*
Internal renvoi, 12-14 *et seq.*
Perfection by filing, recording or registration in Multi-unit States, 12-14 *et seq.*
Public policy and internationally mandatory rules exceptions (limitation) (Art. 11(3)), 11-12
Place of the Relevant Intermediary Approach (PRIMA), *See Conflict of law rules*
Requirements, 2-21 *et seq.*

Permanent Bureau

Drafting initiative, Int-7, Int-14
Information concerning internal choice of law rules in Multi-unit State, 12-17
Initial Report, Int-7

Pledge (of securities or securities account), *See Disposition*

Pooling of Securities, *See also Fungible account*

Rights in, Int-22, Int-23, 2-11
Segregation (no), Int-18

Portfolio of securities, Int-38, Int-42

Practical operation of Convention (review of), 14-1 *et seq.*

Pre-Convention account agreements and securities account, Int-75 *et seq.*, 16-1 *et seq.*

Pre-Convention and post-Convention interests (priority between), Int-74, 15-1 *et seq.*

PRIMA, *See Conflict of law rule (Place of the Relevant Intermediary Approach)*

Primary rule (Art. 4), Int-47 *et seq.*, Int-60 *et seq.*, 4-1 *et seq.*, 4-14 *et seq.*

Amendment of the account agreement, *See Change of applicable law*
Change of applicable law, *See this expression*
Consent, 4-19
Express choice of law, Int-47 *et seq.*, Int-60 *et seq.*, 4-17 *et seq.*, 16-20
Foundation, Int-47 *et seq.*, 4-6 *et seq.*
Multi-unit State, 12-6 *et seq.*, 12-14 *et seq.*, *See also this expression*
Pre-Convention account agreement, 16-1 *et seq.*
Qualifying Office requirement, 4-21 *et seq.*
Account number or bank code (specific identification), 4-38
Office, *See this expression*
Qualifying activity, 4-31 *et seq.*
Time factor, 4-27 *et seq.*
Relevant account agreement, *See Account agreement (relevant)*
Relevant intermediary, *See this expression*
Relevant securities account, 4-43 *et seq.*, *See also Securities account (relevant)*

Priority of competing interests, 1-21, 2-23, 7-22 *et seq.*, 7-25 *et seq.*, 11-12

Priority between pre-Convention and post-Convention interests, Int-74, 15-1 *et seq.*

Promissory notes, 1-2 and *See Securities (definition)*

Protection of rights on change of the applicable law, *See Change of applicable law (Art. 7)*

Public policy (exception), Int-68, 11-1 *et seq.*, 11-6 *et seq.*, 11-12 (limitation)

Purely contractual or otherwise purely personal right, Int-59, 2-7, 2-17, 2-32 *et seq.*

Q

Qualifying Office requirement, *See also Office*

Activity (qualifying), 4-31 *et seq.*
Conditions, Int-61, 4-21 *et seq.*, 4-34 *et seq.*
Definition, 1-45
Identification (means of, incl. account number and bank code), 4-38
(in) Fall-back rules, 1-22, 1-25
Multi-unit State, 12-10 *et seq.*, 12-21 *et seq.* (Option to impose geographically more stringent condition)
(in) Primary rule, Int-61, 1-22, 1-25, 4-21 *et seq.*, 4-34 *et seq.*
Time factor, 1-24, 4-27 *et seq.*

R

Ratification of the Convention, 17-1 *et seq.*

Realisation (of interest in securities) (applicable law), 2-28

Recharacterisation of disposition (risk), 2-19

Redemption (entitlement to) (applicable law), 2-29

Regional Economic Integration Organisations (REIO), Int-77, 18-1 *et seq.*

Registered securities, Int-36, 1-1

Registrar (not an intermediary), 1-33

Regulatory law (no impact on), Int-59, 2-35, 11-12

REIO, *See Regional Economic Integration Organisations (REIO)*

Relevant account agreement, *See Account agreement (relevant)*

Relevant intermediary, Int-30, 1-17 *et seq.*

Account holder's rights (nature of the rights), *See Account holder (rights)*
Book entry, *See this expression*
Definition, 1-17
Description, Int-16 *et seq.*, Int-18 *et seq.*, Int-26 *et seq.*
Disposition in favour of the intermediary, 4-41, 16-8
Legal aspects, Int-19 *et seq.*
Place of business (Art. 5(3)), 5-11, 7-13
Place of incorporation or organisation (Art. 5(2)), 5-7 *et seq.*, 7-13

Relevant securities account, *See Securities account (relevant)*

Renvoi (exclusion), Int-67, 10-1 *et seq.*

Internal renvoi (Multi-unit State), 12-13, 12-14 *et seq.*, 12-21 *et seq.*

Repo, *See Disposition*

Reporting requirement (irrelevant in determining location of securities account), Int-43, 4-26

Reservation, 21-1 *et seq.*, 24-1 *et seq.*

Review of practical operation of Convention, 14-1 *et seq.*

Rights of account holder, *See Account holder (rights)*

S

Sale or other proceeds (entitlement to) (applicable law), 2-29

Scope of the Convention, Int-51 *et seq.*

Article 2(1) issues, *See this expression*
Internationality, Int-53, 3-1 *et seq.*
Intermediated system, *See under Securities held with an intermediary*
Securities, *See this expression*
Securities credited to an account, Int-24, 1-19, 2-15 *et seq.*
Securities held with an intermediary, 1-16, 2-3

Securities

Bearer securities, Int-36, 1-1
Book-entry, *See this expression*
Cash (exclusion), 1-1, 1-5
Certificate, 1-1, 1-2
Debt securities, 1-1, 1-2
Definition, 1-1 *et seq.*
Disposition, *See this expression*
Equity securities, 1-1, 1-2
Holding of securities, *See this expression*
Registered securities, Int-36, 1-1

Securities account (relevant)

Chain of intermediaries, 4-43 *et seq.*
Definition, 1-6 *et seq.*
Disposition in favour of the intermediary, 4-41 *et seq.*, 16-8
Pre-Convention securities account, Int-75, 16-1 *et seq.*

Securities held with an intermediary

Definition, 1-16
Description of the concept, Int-16 *et seq.*

Security certificate, *See Securities (definition)*

Security entitlements, Int-23, 2-11, 12-15, *See also Account holder (rights, nature of the right)*

Security interest, *See Disposition*

Segregation, *See Pooling of Securities*

Shares (listed or not), 1-2 and *See Securities (definition)*

Signature of the Convention, 17-1 *et seq.*, 18-1 *et seq.*

State Party, 1-43, 17-1 *et seq.*

Stock, 1-2 and *See Securities (definition)*

Stock loan, *See Disposition*

Super-PRIMA (rejection of), Int-63, 4-5, 4-11, 4-43 *et seq.*

T

Tax requirement (irrelevant in determining location of securities account), Int-43, 4-26

Territorial Units, See also Multi-uni State

Applicable law in relation to, Int-69 *et seq.*
Internal conflict rules, 3-13

Time

Application of the Convention, 3-11
Qualifying Office test (for), 4-23 *et seq.*

Transfer agent (not an intermediary), 1-33

Transfer of title, *See Disposition*

Transfer from one securities account to another (disposition by), incl. through chain of intermediaries, Int-63, 4-43 *et seq*

Transition dispositions (Arts 15 and 16), Int-74 *et seq.*, 15-1 *et seq.*, 16-1 *et seq.*

U

Units in collective investment schemes, 1-2 and *See Securities (definition)*

Upper-tier attachment, 2-26

W

Warrants, 1-2 and *See Securities (definition)*

Writing or written

Declaration with the Depositary, *See Declaration*
Definition, 1-29
Fall-back rules, 5-4 *et seq.*

Operator Theory: Advances and Applications
Vol. 168